RENEWABLE ENERGY: RESEARCH, DEVELOPMENT AND POLICIES

TECHNICAL CHALLENGES IN THE COMMERCIALIZATION OF TRANSFORMERS FOR SOLAR PHOTOVOLTAIC TECHNOLOGY APPLICATIONS

RENEWABLE ENERGY: RESEARCH, DEVELOPMENT AND POLICIES

Additional books and e-books in this series can be found on Nova's website under the Series tab.

ENERGY SCIENCE, ENGINEERING AND TECHNOLOGY

Additional books and e-books in this series can be found on Nova's website under the Series tab.

RENEWABLE ENERGY: RESEARCH, DEVELOPMENT AND POLICIES

TECHNICAL CHALLENGES IN THE COMMERCIALIZATION OF TRANSFORMERS FOR SOLAR PHOTOVOLTAIC TECHNOLOGY APPLICATIONS

BONGINKOSI A. THANGO,
LEON S. SIKHOSANA,
JACOBUS A. JORDAAN,
AGHA F. NNACHI,
UDOCHUKWU B. AKURU,
ALOYS O. AKUMU
BOLANLE T. ABE

Copyright © 2021 by Nova Science Publishers, Inc.
DOI: https://doi.org/10.52305/JOEP9002

All rights reserved. No part of this book may be reproduced, stored in a retrieval system or transmitted in any form or by any means: electronic, electrostatic, magnetic, tape, mechanical photocopying, recording or otherwise without the written permission of the Publisher.

We have partnered with Copyright Clearance Center to make it easy for you to obtain permissions to reuse content from this publication. Simply navigate to this publication's page on Nova's website and locate the "Get Permission" button below the title description. This button is linked directly to the title's permission page on copyright.com. Alternatively, you can visit copyright.com and search by title, ISBN, or ISSN.

For further questions about using the service on copyright.com, please contact:
Copyright Clearance Center
Phone: +1-(978) 750-8400   Fax: +1-(978) 750-4470   E-mail: info@copyright.com

NOTICE TO THE READER

The Publisher has taken reasonable care in the preparation of this book, but makes no expressed or implied warranty of any kind and assumes no responsibility for any errors or omissions. No liability is assumed for incidental or consequential damages in connection with or arising out of information contained in this book. The Publisher shall not be liable for any special, consequential, or exemplary damages resulting, in whole or in part, from the readers' use of, or reliance upon, this material. Any parts of this book based on government reports are so indicated and copyright is claimed for those parts to the extent applicable to compilations of such works.

Independent verification should be sought for any data, advice or recommendations contained in this book. In addition, no responsibility is assumed by the Publisher for any injury and/or damage to persons or property arising from any methods, products, instructions, ideas or otherwise contained in this publication.

This publication is designed to provide accurate and authoritative information with regard to the subject matter covered herein. It is sold with the clear understanding that the Publisher is not engaged in rendering legal or any other professional services. If legal or any other expert assistance is required, the services of a competent person should be sought. FROM A DECLARATION OF PARTICIPANTS JOINTLY ADOPTED BY A COMMITTEE OF THE AMERICAN BAR ASSOCIATION AND A COMMITTEE OF PUBLISHERS.

Additional color graphics may be available in the e-book version of this book.

Library of Congress Cataloging-in-Publication Data

ISBN: 978-1-68507-214-8

Published by Nova Science Publishers, Inc. † New York

CONTENTS

Introduction		vii
Preface		ix
Chapter 1	Effect of Harmonic Load Currents in Transformer Losses	1
Chapter 2	Stray Losses Analysis Using Finite Element Method	17
Chapter 3	Design Considerations for K-Rated Transformers	37
Chapter 4	Transformer Thermal Performance	55
Chapter 5	Degree of Polymerization of Cellulose Insulation	73
Chapter 6	Transformer Preventive Maintenance of Transformer Health Index through Stray Gassing	91
Chapter 7	A Diagnostic Study of Dissolved Gases in Transformers Based on Fuzzy Logic Approach	109

Chapter 8	Loss Financial Evaluation	**125**
About the Authors		**143**
Index		**149**

INTRODUCTION

It is a great pleasure to introduce this book covering some critical aspects of transformer engineering, particularly renewable energy applications. I would like to thank the co-authors involved in this book for their contribution to all the aspects covered herein.

The renewable energy market has seen a significant increase in South Africa and across the rest of the world in efforts by governments to eradicate greenhouse gases.

Significant developments are necessitated in the design, assessment and diagnostic procedures for transformers in this new era of the renewable energy market to meet some of the stringent operational and economic requirements.

I find that a multitude of these challenges in the design, manufacturer and operation of transformers, to be fulfilled in the rapidly changing energy market settings and engineering solutions, are embellished in this book.

It is hoped that this book will be a vital reference for transformer manufacturers, utility owners and industry researchers involved with the design, improvement and preventative maintenance of transformers for renewable energy applications.

Bonginkosi A. Thango

PREFACE

The presence of harmonic currents as a result of switching action of inverters, intermittent renewable source and resonances triggered by the connection of the solar photovoltaic plant to the electrical network have spread the realization of the prospective rapid depletion of transformer's intended service life owing to increased service losses and temperature rise in the active part components. On the grounds of the supposed prospect of reduced transformer service lifetime under harmonic conditions, manufacturers have a growing interest in exploring transformer design procedures that enable transformers to operate reliably under harmonic conditions.

Chapter 1

EFFECT OF HARMONIC LOAD CURRENTS IN TRANSFORMER LOSSES

Various utility owners have reported the untimely failures of transformers in solar photovoltaic (PV) technologies despite local manufactures adopting the design philosophies recommended by the existing standards during the order design stage. The problem arises on the reality that the existing standards are solely adequate for regular distribution transformers that are intended to be in-service to a vertically integrated utility in the case of South Africa. These transformers will pass all the Factory Acceptance Tests (FATs) only to have a shorter in-service lifetime and to exhibit oil stray gassing patterns. These accentuate the need to perform comprehensive studies in relation to analyzing the effects of harmonic load currents in solar PV transformers, specifically on the winding Eddy losses and structural stray losses. The aforementioned is crucial in examining the capability of these transformers under the stringent operational requirements of solar PVs.

This chapter will investigate the capability of an oil-filled solar PV transformer case study susceptible to a harmonically contaminated

Point of Common Coupling (PCC) on a grid-connected solar PV plant. To accomplish this objective, the harmonic loss factor to account for additional winding Eddy losses of the transformer while in service owing to harmonic currents is evaluated. In addition, the effect of the skin effect on the winding conductors under harmonic conditions is considered. Finally, an improved harmonic loss Factor for winding Eddy losses is recommended. The investigated case scenario and results are then used to examine the transformer's capability when supplying distorted load current for existing and planned solar PV development.

1. INTRODUCTION

The national electric grid of South Africa has gone through significant adjustments over the past decade. The government has committed to investing in alternative solutions to conventional coal energy production technology by exploring renewable energy sources, particularly solar, wind and hydro. The unyielding commitment to alleviate the release of greenhouse gases has led to an enormous growth of solar PVs in South Africa. On account of their excellent location for this application, the provinces leading in the photovoltaic power generation include Northern Cape, Western Cape and Free State [1]. The present photovoltaic power generation capacity in South Africa is about 1,274MW [2]. Furthermore, determinations have been indicated by the Integrated Resource Plan (IRP) on solar PV energy that expects a generation capacity of 8,500MW by 2030 [3].

Some of the coal power plants in South Africa are in the process of being decommissioned. In the same vein as many African countries, South Africa intends to produce renewable energy at cost-competitive and low cost tariffs [4]. The South African government has invested ZAR193-billion within the last decade to deploy new solar PV s [5-7]. In 2015, The IRP report was released, in which some ministerial determinations to procure 6,328MW of solar PVs was proposed while recognizing similar opportunities in 2016 [5-7]. By the end of 2017, there were 92 solar PV installed in South Africa [5-7]. The Department of Energy (DOE) plans to deploy more solar PV s in the next decade with a projected contribution of 14.9% of South Africa's total generation capacity [8].

Independent Power Producers (IPP'S) are predicting a huge dependency on photovoltaic power generation across the African continent in the near future. A solar PV transformer is installed across a fleet of inverters driven by a cluster of parallel solar modules. The primary duty of this transformer is to step up the voltage output of inverters from the power generation system to a desired voltage level of the medium voltage (MV) network. Untimely failures of solar PV transformers have transpired at a worrisome rate, causing IPP's and transformer manufactures to conduct failure investigations to identify the root causes of reported failures. In any event, most utilities have gone through solemn challenges from the lookout of power system harmonics studies when integrating solar PVs into the existing electric grid.

Non-linear loads and inverters initiated harmonics and distortion are foreseen to be one of the main causes behind the solar PV transformers failures. The inverters inputs driven by solar modules are normally switched on and off various times throughout the day due to variations in solar radiation, resulting in enormous fluctuations in the energy generated. The resultant distorted harmonic current can produce high losses, resulting in hotspot temperature rise and thus thermal ageing-related failures.

The suggested customary action in the estimation of harmonics when supplying non-linear loads include the assumption that winding Eddy losses in the winding conductors increase with the square of the harmonic order, but fail to provide accurate result about the Eddy losses of large winding conductors at high frequencies experienced by the solar PV transformers. This accentuates the importance of the correction of the winding Eddy loss and consequently the thermal requirements for solar PV transformers.

2. TRANSFORMER HARMONIC LOSSES

2.1. H^2 Method

In view of the fact that the biggest interest regarding a solar PV transformer operating below harmonics and distortion condition is the hotspot temperature rise of the windings, it is serviceable to contemplate on the loss amount per unit in the windings on a per-unit basis. The rated current is considered as the base and the copper loss ratio at rated current is the base loss ratio. The total load loss on a per-unit basis is as shown in Eq. (1) [9-11].

$$P_{LL(p.u)} = 1 + P_{EC-R}(p.u) + P_{SSL-R}(p.u) \tag{1}$$

Provided the winding Eddy losses at rated condition, the service winding Eddy losses under harmonic conditions can be expressed as shown in Eq. (2) [9]-[11]. This loss present the efficient mean temperature rise due to harmonic load current.

$$P_{WEC} = P_{WE\_R} \times \sum_{h=1}^{h=h_{max}} \left(\frac{I_h}{I_R}\right)^2 \times h^2 \tag{2}$$

It is expedient to determine a unit value that could be employed to specify the robustness of a solar PV transformer when supplying power to the solar PV plant's loads. The harmonic loss factor is applied to the winding Eddy loss at measured load current and fundamental frequency to ascertain the service winding Eddy loss under harmonic conditions. This expression is shown in Eq. (3) [9]-[19].

$$F_{WEC} = \frac{\sum_{h=1}^{h=h_{max}} \left(\frac{I_h}{I_R}\right)^2 \times h^2}{\sum_{h=1}^{h=h_{max}} \left(\frac{I_h}{I_R}\right)^2} \tag{3}$$

Eq. (3) allows the harmonic loss factor to be determined in terms of the ratio of the total harmonic load current to the r.m.s fundamental load current. In Eq. (4), the structural stray losses (P_{SSL}) causing temperature rise on the solar PV transformer oil is expressed. This loss is established using the same philosophy as the winding Eddy losses above. However, the losses due to tank walls, flitch plates core clamps, frames and bus-bars are proportional to the product of the load current and harmonic order to the exponent 0.8 as shown in Eq. (4) [9-11].

$$P_{SSL} = P_{SSL\_R} \times \sum_{h=1}^{h=h_{max}} \left(\frac{I_h}{I_R}\right)^2 \times h^2 \tag{4}$$

The harmonic loss factor corresponding for the structural stray losses is expressed as shown in Eq. (5).

$$F_{SSL} = \frac{\sum_{h=1}^{h=h_{max}} \left(\frac{I_h}{I_R}\right)^2 \times h^{0.8}}{\sum_{h=1}^{h=h_{max}} \left(\frac{I_h}{I_R}\right)^2} \tag{5}$$

The per-unit winding Eddy loss in the region with the highest potential of hotspot temperature can be explained for the fundamental frequency operation at rated current by the solar PV transformer manufacture using Eq. (6) [9-19].

$$P_{EC(p.u)} = I^2{}_{rms(p.u)} \times (1 + F_{PEC} \times P_{EC-R}(p.u)) \tag{6}$$

In any event, Eq. (6) does not take account of the structural stray losses due to the fact that the stray losses are not present in the winding conductors.

2.2. Maximum Load Current

The resultant maximum allowable non-sinusoidal load current can be computed as shown in Eq. (7) [9-19]. This current suggest the capability of the transformer of its fundamental load current.

$$I_{max}(pu) = \sqrt{\frac{P_{LL-R}(p.u)}{1+F_{PEC} \times P_{EC-R}(p.u)}} \tag{7}$$

The corresponding maximum permissible non-sinusoidal current of the solar PV transformer for a specific harmonic profile will then be expressed as shown in Eq. (8) [9-11].

$$I_{max} = I_{max(p.u)} \times I_{rated} \tag{8}$$

2.3. Improved Harmonic Loss Factor

The skin effect is a critical factor that needs to be taken into consideration when determining the service losses of a solar PV transformer. At high harmonic order, the skin effect restricts the permeation of magnetic fields upon the conductors. The assumption that the winding losses increase with the square of the harmonic current is untrue at high frequencies. An improved harmonic loss factor that will answer for the additional losses when a transformer is supplying distorted load current is proposed based on a function used in 1966 by Laminar and Lehton [20]. In Eq. (9), a function that considers the skin depth related to winding conductor dimensions is presented.

$$F(p) = \frac{3}{p} \frac{\sinh p - \sin p}{\cosh p + \cos p} \tag{9}$$

The relation between the skin depth and winding conductor is expressed in Eq. (10). Here, T is the axial and radial thickness of the conductor perpendicular to the direction of the magnetic flux density.

$$p = \frac{T}{\delta} \tag{10}$$

The depth of penetration of the magnetic flux density at fundamental can be expressed as shown in Eq. (11).

$$\delta_R = \sqrt{\frac{\rho}{\pi \mu_0 f}} \tag{11}$$

The skin depth for copper conductors at the fundamental frequency of 50Hz and 75°C is 10.63 millimeters. Under harmonic conditions, the depth of penetration is expressed as shown in Eq. (12).

$$p_h = \frac{H_c}{\delta_R} = \delta_R \times \sqrt{h} \tag{12}$$

The improved winding Eddy losses on account of harmonic current is then expressed as shown in Eq. (13).

$$P_{WEC} = P_{WE\_R} \times \sum_{h=1}^{h=h_{max}} \left(\frac{F(p_h)}{F(p_r)}\right) \left(\frac{I_h}{I_R}\right)^2 \times h^2 \tag{13}$$

The corresponding improved harmonic loss factor standardized on the basis of the r.m.s current is shown in Eq. (14).

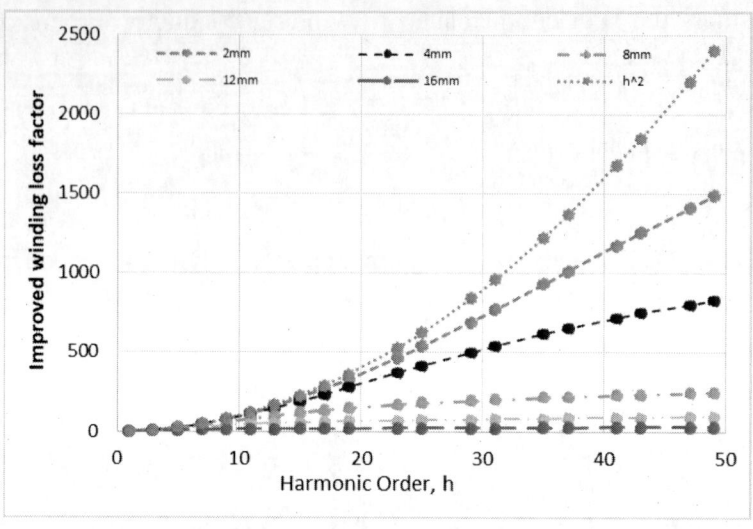

Figure 1. Improved harmonic loss factor for various winding conductor dimensions.

The square of the harmonic order, h^2, is included in Figure 1 for comparative purposes. It is observed that for conductors with dimensions less than 3 millimeters glide towards the function h^2. Additionally, the influence of skin effect is significantly low for conductors with dimensions less than 3 millimeters and the method presented in section 2.3 is reasonably accurate. However, as a result of the ignored skin effect at harmonic order greater than the 6th order and with large winding conductor dimension, the results are inaccurate.

$$F_{WEC} = \frac{\sum_{h=1}^{h=h_{max}} \left(\frac{F(p_h)}{F(p_r)}\right)\left(\frac{I_h}{I_R}\right)^2 \times h^2}{\sum_{h=1}^{h=h_{max}} \left(\frac{I_h}{I_R}\right)^2} \tag{14}$$

The significance of the improved harmonic loss factor is shown in Figure 1, where the plots are defined as a function of the harmonic order for various winding conductor dimensions.

3. CASE STUDY: AN OIL-FILLED SOLAR PV TRANSFORMER

In this section, an oil-filled solar PV transformer with a rated current of 131.2A is studied. The rated losses at fundamental frequency for this transformer is shown in Table 1. The unit has rectangular copper conductor dimensions of 2x13.95mm on the low voltage disc winding.

The transformer mentioned above is intended to operate in a solar PV plant with the harmonic profile shown in Figure 2. While the transformer is in service, it will see a significant harmonic current of 0.45 p.u for loads with a 3rd harmonic order.

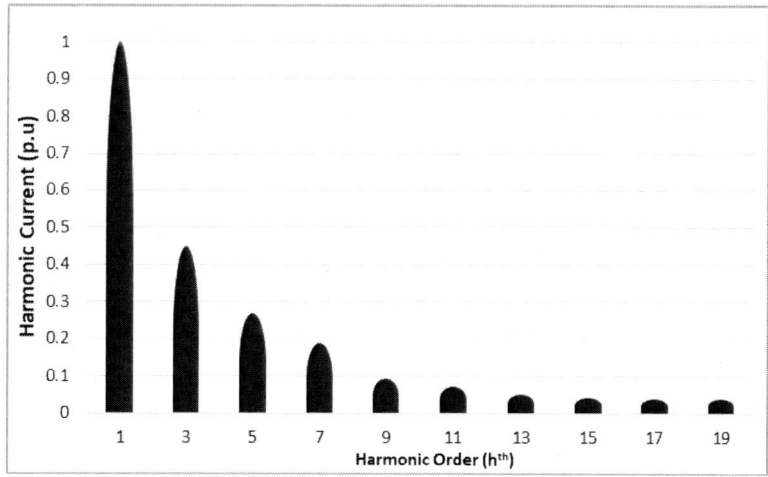

Figure 2. Solar PV plant harmonic profile.

Table 1. Transformer Losses under Fundamental Condition

Type of loss	Losses (W)
No-load	3 938
I²R	32 275
Eddy Losses	1 195
Total stray	1 305
Total losses	38 713

3.1. Transformer Harmonic Loss and Capability

In this sub-section, the estimation of the service losses under the given harmonic spectrum and the capability of the transformer is defined as shown in Table 2. A per-unit r.m.s load current of 1.154 is attained from the harmonic spectrum. The partitioning of the losses for the transformer under study at the given harmonic profile using the h^2 assumption is tabulated as shown in Table 1.

Table 2. Computed loss under the supplied harmonic spectrum

Type	Rated loss (W)	Load Losses (W)	Harmonic Multiplier	Service Losses (W)
No-load	3 938	3 938	-	3 938
I²R	32 275	43 009	-	43 009
WEL	1 195	1 592	7.19	11 444
OSL	1 305	1 739	1.56	2 707
Total losses	38 713	50 278	-	61 098

The computation of the improved harmonic losses resulting from the harmonic profile under study is tabulated as shown in Table 3.

Table 3. Improved Loss Distribution under the Supplied Harmonic Spectrum

Type	Rated loss (W)	Load Losses (W)	Harmonic Multiplier	Service Losses (W)
No-load	3 938	3 938	-	3 938
I²R	32 275	43 009	-	43 009
WEL	1 195	1 592	1.464	2 332
OSL	1 305	1 739	1.56	2 707
Total losses	38 713	50 278	-	51 985

The per-unit winding Eddy losses under the given harmonic profile in the region of highest Eddy current is calculated using Eq. (6) and the h² method.

$$P_{EC(p.u)} = I^2_{rms} \times (1 + F_{PEC} \times P_{EC-R}(p.u)) = 1.154^2 \times (1 + 7.19 \times 0.037) = 1.69 \, p.u$$

For the case of the improved harmonic loss factor, the winding Eddy losses are calculated as:

$$P_{EC(p.u)} = I^2_{rms} \times (1 + F_{PEC} \times P_{EC-R}(p.u)) = 1.154^2 \times (1 + 1.464 \times 0.037) = 1.33 \, p.u$$

The above has a percentage error of about 21.016% is observed for the winding Eddy losses.

3.2. Maximum Allowable Non-Sinusoidal Load Current

For the given harmonic profile of the solar PV plant, the maximum allowable non-sinusoidal load current in per-unit is treated by employing Eq. (7) as follows:

$$I_{max}(p.u) = \sqrt{\frac{P_{LL-R}(p.u)}{1+F_{PEC} \times P_{EC-R}(p.u)}} = \sqrt{\frac{1.077}{1+7.19 \times 0.037}} = 0.9 \, p.u$$

The corresponding maximum current is calculated in accordance with Eq. (8) as follows:

$$I_{max} = I_{max(p.u)} \times I_{rated} = 0.9 \times 131.22 = 118.15 \, A$$

This results suggest that the suitable maximum allowable load current is 118.15A and the solar PV's transformer capacity is now about 90% of its sinusoidal load current capability. By considering the impact of the skin effect and treating the service Eddy losses using the improved harmonic loss factor, the maximum non-sinusoidal load current is as follows:

$$I_{max}(p.u) = \sqrt{\frac{P_{LL-R}(p.u)}{1+F_{PEC} \times P_{EC-R}(p.u)}} = \sqrt{\frac{1.077}{1+1.464 \times 0.037}} = 0.982 \, p.u$$

Similarly, corresponding maximum current is calculated in accordance with Eq. (8) as follows:

$$I_{max} = I_{max(p.u)} \times I_{rated} = 0.982 \times 131.22 = 128.868 \, A$$

This result indicates that the maximum allowable current is reduced to 128.868A and the transformer capacity is now approximately 98.2% of its sinusoidal load current capability. This method enables for a 9.07% increase of the current.

CONCLUSION

This chapter presents a harmonic spectrum to study the harmonic and distortion effects on the point of common coupling linked to an oil-filled transformer in a solar PV plant. The results reveal that the assumption that the winding Eddy losses increase with the square of the harmonic current may cause incorrect estimation of these losses at high harmonic order due to the limitation of the magnetic field to pierce through the winding conductors. The improved harmonic factor was observed to have a significantly lower value than the h2 method at high harmonic order due to this effect. Additionally, the proposed method supplied an increase in the maximum allowable load current of about 9.07%.

At large, this chapter serves as a guide for electrical designers during the design stage provided that the planned solar PV plant owners provide a complete harmonic profile to avoid under-designing the transformer, in which it then fails to meet its service technical including cooling and thermal requirements. Conversely, if the solar PV transformer is over-designed, the initial purchase will be unnecessarily expensive for the utility owners, with the transformer only utilizing a fraction of the designed capacity during its service life.

REFERENCES

[1] Power Africa, *Energy Transactions and Projects*, 2018. Online. Available: https://www.usaid.gov/powerafrica/south-africa-power-africa-transactions.

[2] National Energy Regulator of South Africa (NERSA), *Monitoring Renewable energy Performance of Power PA*, 2018.

[3] Electric Power Research Institute (EPRI), "Power Generation Technology Data for Integrated Resource Plan of South Africa," *Technical Update*, April 2017.

[4] *Global Solar Atlas*, "The World Bank, Solar resource data: Solargis," 2017.

[5] Lilley R., *An invitation to invest in South Africa's energy sector*, EE Publishers, September 2018.

[6] South African Photovoltaic Industry Association (SAPVIA), *South Africa Photovoltaic Industry Association welcomes long-awaited draft Integrated Resource Plan*, August 2018.

[7] Ratshomo K., "South African Energy Sector Report," *Department of Energy*, 2018.

[8] Modise D., V. Mahotas, *South African Energy Sector Presentation*, Department of Energy. [Online]. Available: https://www.usea.org/sites/default/files/event-file/497/South _Africa_Country_Presentation.pdf.

[9] Thango B. A., J. A. Jordaan and A. F. Nnachi, "Effects of Current Harmonics on Maximum Loading Capability for Solar Power Plant Transformers," *2020 International SAUPEC/ RobMech/PRASA Conference*, Cape Town, South Africa, 2020, pp. 1-5, doi: 10.1109/SAUPEC/RobMech/PRASA48453. 2020.9041101.

[10] Arslan E., M. E. Balci and M. H. Hocaoglu, "An analysis into the effect of voltage harmonics on the maximum loading capability of transformers," *2014 16th International Conference on Harmonics and Quality of Power (ICHQP)*, Bucharest, Romania, 2014, pp. 616-620, doi: 10.1109/ICHQP. 2014.6842925.

[11] Gado A., H. A. Gad, S. Radwan, "Effect of Types of Loads in Rating of Transformers Supplying Harmonic-rich Loads," *21st International Conference on Electricity Distribution*, Frankfurt, 6-9 June 2011.

[12] Pejovskia D., K. Najdenkoskib, M. Digalovski, "Impact of different harmonic loads on distribution transformers," *4th International Colloquium Transformer Research and Asset Management*, Pg. 76–87, Procedia Engineering 202, 2017.

[13] Said D. M., K. M. Nor, "Effects of Harmonics on Distribution Transformers," *2008 Australasian Universities Power Engineering Conference*, University of New South Wales, Sydney, Australia 14-17.12.2008.

[14] Ozerdem O. C, A. Al-Barrawi, S. Biricik, "Measurement and comparison analysis of harmonic losses in three phase transformers," *International Journal on Technical and Physical Problem in Engineering*, Vol.5 Issue 14, March 2013, pp. 114-118.

[15] Das B. P. and Z. Radakovic, "Is Transformer kVA Derating Always Required Under Harmonics? A Manufacturer's Perspective," in *IEEE Transactions on Power Delivery*, vol. 33, no. 6, pp. 2693-2699, Dec. 2018, doi: 10.1109/TPWRD.2018.2815901.

[16] Yildirim D. and E. F. Fuchs, "Measured transformer derating and comparison with harmonic loss factor (F/sub HL/) approach," in *IEEE Transactions on Power Delivery*, vol. 15, no. 1, pp. 186-191, Jan. 2000, doi: 10.1109/61.847249.

[17] *IEEE Recommended Practice for Establishing Liquid Immersed and Dry-Type Power and Distribution Transformer Capability when Supplying Nonsinusoidal Load Currents*, IEEE C57.110-2018, June 2018.

[18] Faiz J., M. Ghazizadeh, H. Oraee, Derating of transformers under non-linear load current and non-sinusoidal voltage – an overview, *The Institution of Engineering and Technology*, July 2015.

[19] Gupta A., A. Soni, "The Performance Analysis of Distribution Transformer under Domestic Harmonics Load," *International Journal of Innovative Research in Electrical, Electronics, Instrumentation and Control Engineering*, November 2016.

[20] Laminar J and M. Stafl, "Eddy Currents," *London Lliffe*, 1966.

Chapter 2

STRAY LOSSES ANALYSIS USING FINITE ELEMENT METHOD

The integration of classical transformer design procedures with the staggering speed and coherent versatility of modernistic computational methods to evaluate and manage stray losses in transformers incite better understanding into the transformer design philosophy, especially for renewable energy (RE) application. This is on account of tools such as Finite Element Method (FEM) that can perform several unwieldy and iterative computations in a judicious and stepwise approach yielding transformer designs that meet the stringent technical specification of REs.

In the present chapter, the key objective is to carry out the computation of the winding Eddy loss, which fosters mapping the hotspot temperature in the transformer windings. To enable the computation of dispersed winding Eddy loss, FEM based Ansys Maxwell is employed to treat the magnetic flux density calculations at the hub of individual winding conductors. The vector decomposition of the flux density gives winding Eddy loss as a result of axial and radial magnetic flux leakage components.

1. INTRODUCTION

In recent decades, a surge in the emergence of research toward the utilization of modernistic computational methods to evaluate and manage stray losses in transformers has been witnessed. These losses can be significantly higher in transformers intended to be of service to renewable energy applications due to the irregularity of renewable energy sources and related arduous operating conditions. A daunting task for transformer designers in this day and age is the capitalization of significantly higher guaranteed loss requirements, competitive transformer pricing, and superlative performance. The advancement of computational power enables the utilization of tools such as the Finite Element Method (FEM) to grant opportunities to improve transformer designs and performance. The introduction of FEM into the transformer challenges and design philosophy in a renewable application has ceded new horizons and principles in addition to economizing the engineering labour hours.

In the past, analytical-based methods such as Rabin's method have been employed to estimate the leakage fields and stray losses [1-2]. This method considers the uniform flux in the distance between the windings. A drawback of these methods is that they fail to account for the fringing effect and non-homogeneity of the magnetic flux leakage. Authors in [3, 4, 5] and [6] have also presented some work based upon analytical methods.

Earlier work on the estimation of the stray losses by numerical methods is by Stoll [7] and Girgis [8]. In the early 2000s, adjustments in the transformer design philosophies towards the application of Finite Element Method (FEM) to compute the winding Eddy current and structural parts losses become known in work published by authors suchlike Kulkarni [9], Del Vecchio [10] and many authors in [11, 12, 13, 14, 15, 16] and [17]. The spread of FEM into the RE applications has been recently witnessed to emerge in publications [18, 19, 20].

In order to ascertain a comprehensive understating of the stray losses in transformers, FEM modelling has been conducted in this chapter for the computation of these losses. The second part of this chapter is focused on the measured losses. Finally, the hotspot temperature as a result of the winding stray losses is presented.

2. LOAD-LOSSES

Many researchers [21–23] have made an effort to provide a comprehensive theoretical and practical analysis of the Eddy current to treat composite engineering problems. These publications provide a premise to transformer manufactures for pragmatic computation of the stray load losses during the transformer design stage.

2.1. Copper Losses

The IEEE Std. C57.110-2018 classify the transformer load losses into copper losses and winding stray losses. The copper loss is computed by using the measured resistance during the factory load loss test and the load current. The stray loss is then obtained by subtracting the copper loss from the measured load loss. Some work on the improved estimation of the copper loss has been witnessed in [24] and [25]. In [24], Dimitrakakis et al. presented an expression to compute this loss under harmonic conditions by employing finite element analysis (FEA). The results are compared with measurement results, yielding improved accuracy and insight on the shortcomings of analytical formulations suchlike Butterworth's formula, Dowell's model and Ferreira's model. In [25], Kubota presents some work based on the application of magnetic flux path control engineering. The methodology takes into account the increase of copper loss under harmonic conditions as a result of skin and proximity effect. The results indicate that the reduction of hotspots can be suppressed by modifying the transformer winding structure to yield a reduction of about 19.9% more than prior to modification.

2.2. Winding Eddy Loss

A prevalent numerical method for the computation of winding Eddy loss among transformer manufactures is the axisymmetric 2 dimensional (2D) Finite Element Method modelling as shown in Figure 3.

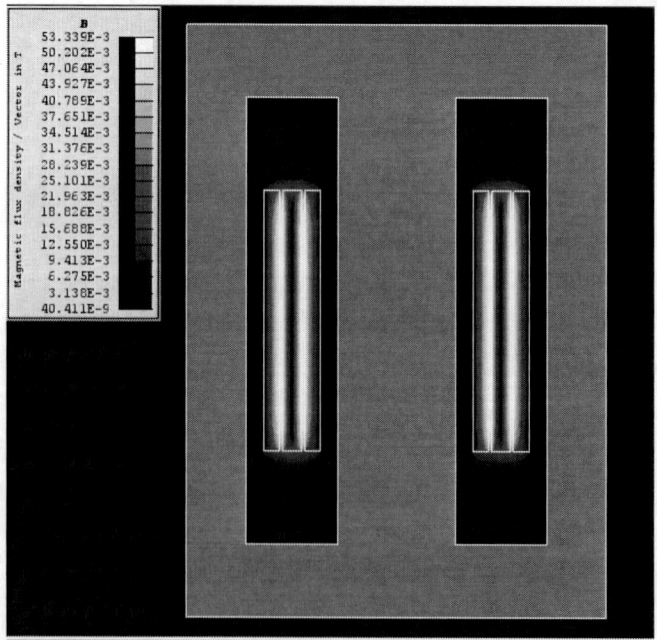

Figure 3. Flux density distribution in the windings.

The realization of the flux density dispersal can be wielded in the selection of the suitable conductor dimensions. The Continuously Transposed Conductor (CTC), which is composed of a bundle of enamelled rectangular copper wires, with Polyvinyl Formal Flat (PVF) enamel, which is stranded to form a rectangular rope, can be used to minimize the guaranteed winding Eddy losses that will be seen by the high current winding during service. In order to evaluate the hottest spot temperature location along the winding height, an accurate estimation of the winding Eddy loss on each winding discs must be realized. The hottest spot region is concentrated at the apex of the winding on account of the inclined radial magnetic flux leakage. Within the last decade, studies is [26, 27, 28, 29, 30] and [31] have been carried out with a special focus on the 2D and 3D computation of the leakage flux in transformers during service.

2.3. Tank Wall Loss

The stray losses induced in the tank walls account for the leakage flux evading the transformer core. The mitigation of this loss is pivotal to transformer manufacturers as they directly impact the unit's performance during the intended service lifetime. The placement of magnetic shunts on the tank walls is common in mitigating stray losses and eradicating potential hotspot regions in the tank walls. Magnetic shield institute a low reluctance trajectory for the leakage flux and hinder it from contact with the tank walls.

In [32], Moghaddami et al. proposed a horizontal arrangement of the magnetic shunts over the conventional vertical arrangement as an economical method to reduce the stray losses in tank walls. Here, Moghaddami et al. employ 3D Finite Element Analysis (FEA) to treat the losses in tank walls. To account for the dispersal of the magnetic fields upon the horizontal tank shunts, Fourier series expansion is employed. The proposed method yield results that indicate a reduction in the size of the horizontal magnetic shunt over conventional method. Earlier investigations on the use of horizontal magnetic tank shunts has also been studied by authors in [33] and [34]. Various magnetic tank shunts have also been proposed by authors in [35, 36] and [37] by means of FEM analysis. At large, the authors highlight that the computation of stray fields in tank walls especially with magnetic tank shunts, 3D modelling is more appropriate to account for the arrangement of the shunts.

2.4. Core Clamp and Frame Loss

Core clamps and frames firmly fastening the top and bottom yokes and a pillar to windings, are nearby the winding stray fields. The flux leakage leaving the inner boundary of the winding conductors impinge upon structural components suchlike the core, core clamp and frames and flitch plates. As a result of spacious surface area of the core clamps and frames, the hotspots can easily be generated. FEM is also quite common amongst transformer manufactures to compute the frames and core clamps. These structural components are made up of mild steel material. Reduction of the Eddy currents is predominantly by aluminium shielding for the transformer windings. Studies based on the evaluation of core clamps and frame losses have been covered by the authors in [38, 39], and [40]. In [40], Mokkapaty et al. introduced a study for reducing the losses in core clamps and frames by applying an electrical magnetic shunt to guard these structural components from the stray flux emanating from the high current leads connected to the low voltage bushing side of the transformer. The evaluation of the losses in these components is treated by using the impedance boundary of the electrical magnetic shunt.

2.5. Flitch Plate Loss

In spite of the fact that the losses in flitch plates do not contribute a considerable amount of the transformer stray losses, the local rise in temperature is significant on account of excessive incident flux density and deficient cooling conditions. If the suitable choice of flitch plate and material is not selected, the loss in this structural component may surge and subsequently to an abnormal temperature rise. There is barely any research available on the flitch plates and evaluation of the losses. Various shortcomings of analytical methods to evaluate the flitch plate losses by transformer manufacturers are prevalently overcome by the use of FEM. In Figure 4, a FEM model of the core clamps and flitch plates embedded in the transformer core are presented. In practice, slitting and lamination are common in reducing the Eddy currents in flitch plates.

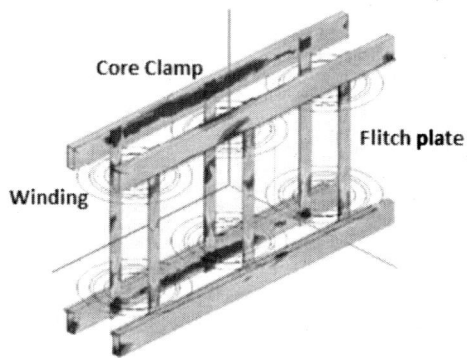

Figure 4. Flux density distribution in the windings.

2.6. Transformer Loading

The firmly established thermal model used for the estimation of the transformer loading is described in the South African National Standard (SANS):60076-2 loading guide [41] for oil-immersed transformers. The proposed explanation made on the basis of temperature distribution is demonstrated in Figure 5.

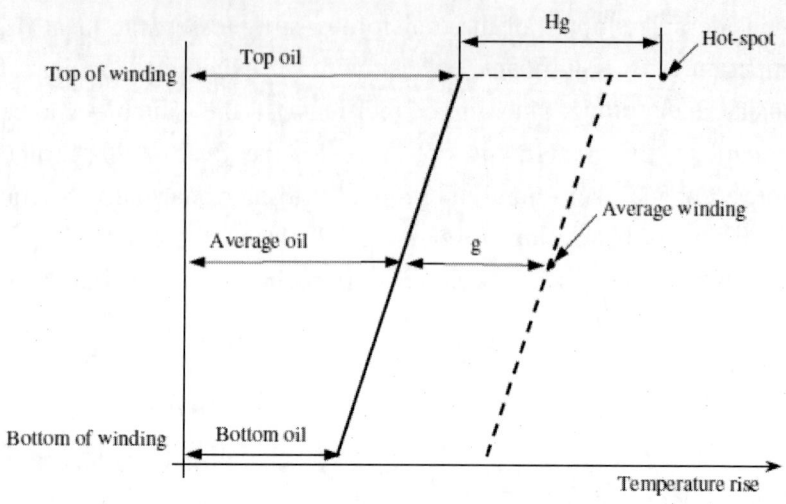

Figure 5. Transformer thermal loading [41].

A linear progression of the oil and winding temperature rise is observed from the bottom to the top of the winding conductors. The winding temperature rise has an invariant temperature difference, g. additionally, the hottest spot temperature at the apex of the winding is significantly higher than the mean winding temperature rise. The hotspot temperature factor, which reflects the nonlinear dispersal of the temperature rise is denoted by, Hg, and is computed as the difference between the hottest spot and top-oil temperature. The recommenced maximum rated thermal requirements for oil-filled transformers are tabulated in Table 4.

At large, the hottest spot temperature (θ_H) under any loading (L) conditions is the arithmetic sum of the ambient temperature (θ_A), top-oil temperature (θ_{TO}) and the temperature gradient (θ_g).

$$\theta_H = \Delta\theta_A + \Delta\theta_{TO} + \Delta\theta_g \tag{1}$$

Table 4. Temperature limits

Temperature	Value
Top Oil (°C)	65
Mean Winding (°C)	60
Hotspot (°C)	78

In addition, the top-oil temperature rise is computed as expressed in eq. (2) [52].

$$\Delta\theta_{TO} = \Delta\theta_{TO\_Rated} \left(\frac{1 + \left(\frac{LL_{Rated}}{NLL_{Rated}}\right) \times L^2}{1 + \left(\frac{LL_{Rated}}{NLL_{Rated}}\right)} \right)^n \quad (2)$$

The exponent, n, is derived through statistical data of the cooling methods employed in various oil-filled transformers and is recommended as 0.9 for ONAN cooling modes in the (SANS):60076-2 [41]. Finally, the temperature gradient is computed by further considering the mean winding temperature rise ($\Delta\theta_{winding(mean)}$), mean top-oil temperature rise ($\Delta\theta_{top-oil(mean)}$) and the exponent, m, also derived based on the transformer cooling method. In the case of ONAN cooling, this exponent is recommended as 0.8 [41].

$$\Delta\theta_g = HSF \times \left(\frac{\Delta\theta_{winding(mean)}}{\Delta\theta_{top-oil(mean)}} \right) \times L^{2m} \quad (3)$$

During the factory acceptance tests (FATs), the temperature rise test [41] does not account for the increased transformer losses as a result of Eddy currents in the apex of the winding ends during service. These additional losses will be taken into account by the HSF. Statistical data provided by the CIGRE Working Group 12-09 [41] suggest the HSF value ranges between 1 and 1.5. In practice, these values can change significantly conditional on the design type. At the design stage, the HSF must be estimated accurately to avoid penalties for understating the guaranteed loading conditions.

3. CASE STUDY

This section presents a 1260kVA, oil-immersed, ONAN cooling transformer to evaluate the load and stray load losses. The low voltage winding of the transformer is a Helical winding, consisting of 3 transposed cables in parallel in each turn and a strand dimension of 2.05×5.2 mm. The high voltage winding is a Disc winding consisting of 25 strands in parallel in each disc and a strand dimension of 2.35×13.5 mm. A post-process of the FEM simulation showing the dispersal of the axial and radial magnetic flux leakage along the windings is presented in Figure 6. In the apex of the windings, the winding conductors are susceptible to a disposed of axial and radial magnetic flux leakage. The winding Eddy losses computed in this section is constituted by these components of the flux leakage.

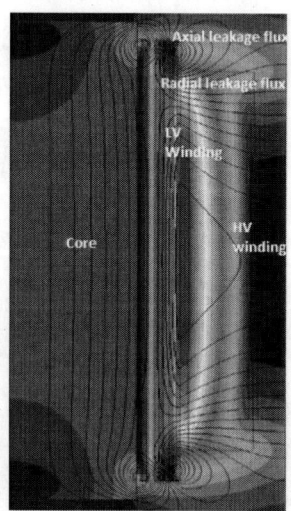

Figure 6. Transformer 2D FEM model.

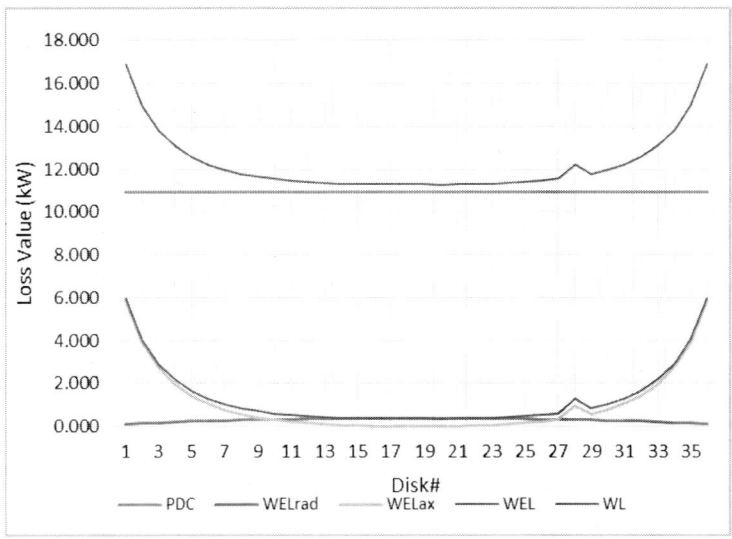

Figure 7. Winding eddy loss for individual disc.

The winding Eddy loss results computed by FEM across individual winding discs at an operational temperature of 75°C are shown in Figure 7. Each winding disc is prone to the local flux density axial and radial components, which trigger the generation of Eddy currents. The radial winding Eddy loss component materialize on the vertical side (height) of the winding conductors while the axial winding Eddy loss on the horizontal (width) side of the winding disc dimensions. Notably, the magnitude of these losses is reliant on the winding dimensions and the magnetic field strength, which is necessitated to produce the flux density components within the copper conductor and other metallic part materials on a per unit length basis.

Furthermore, the load loss results are used to treat the estimation of the hottest spot region on the winding conductors. The ratio of the maximum winding Eddy loss to the mean winding Eddy loss is used to compute the hotspot factor, H, required to predict these temperature as expressed in eq. (4).

$$H = \left(\frac{WEL_{max}}{WEL_{average}}\right)^{0.8} \tag{4}$$

At fundamental frequency, the hottest spot factor is estimated to be 1,374 and the corresponding thermal requirements of the transformer under study are presented in Table 5.

Table 5. Measured transformer losses

Temperature	Model	Measured
Top Oil	61.5	59.8
Mean Winding	57.8	55.6

Table 6. Measured transformer losses

Loss Parameter	Model	Measured
PDC	13,90	14,04
PLL	16,41	16,58
WEL+ OSL	2,51	2,54

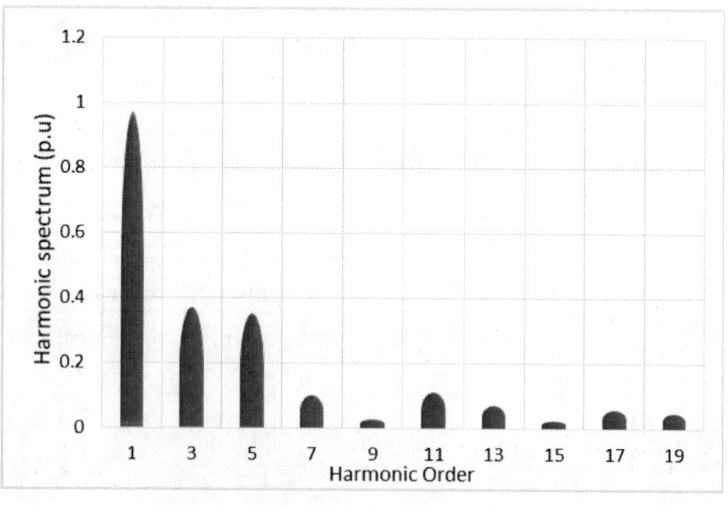

Figure 8. Percentage distribution of the load loss.

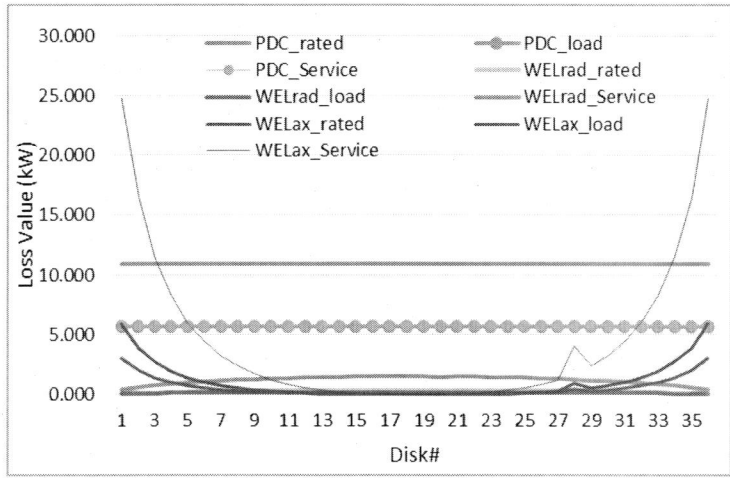

Figure 9. Winding Eddy Loss Distribution.

The measured transformer load losses during Factory Acceptance Tests (FATs) at the transformers principal tap position is tabulated in Table 6.

The harmonic profile considered in this study to account for the additional losses that will be seen by the transformer during service is presented in Figure 8.

Based on the supplied harmonic spectrum, the load loss and service loss components are computed in Figure 9. Some insight on the distribution of the various losses indicates a bath tub characteristic with a concentration of the winding Eddy losses towards the winding ends.

The service axial winding Eddy loss under the supplied harmonic spectrum is evidently higher than the service axial winding Eddy loss.

CONCLUSION

In the new dawn of renewable energy and arduous operating conditions linked to high transformer design and manufacturing costs, unerring and optimal evaluation of stray losses by modernistic computational methods, suchlike FEM cedes techno-economic superiority among other competing transformer offers. In the current chapter, a FEM model is developed and employed to evaluate the transformer stray losses. The information obtained on the flux density is used taking into account the winding conductor dimensions to evaluate the winding Eddy losses in the studied design. The developed model can determine the winding Eddy losses in individual discs to map out the hotspot factor and, consequently, the hotspot temperature.

The chapter further gave an in-depth insight into the literature review of the various transformer stray losses. In future work, authors will extend the study to investigate the management of stray losses by means of procedures suchlike magnetic shielding.

REFERENCES

[1] Hlatshwayo M. C, *The computation of winding eddy losses in power transformers using analytical and numerical methods*, http://hdl.handle.net/10539/12891.

[2] Pasquotto de Lima P. A. and S. H Lopes Cabral, *A Practical Application of the Rabin´s Method for Inductance Calculation in Power Transformer Design*, 10° CBMag – Congresso Brasileiro de Eletromagnetismo.

[3] Leal A. G., J. A. Jardini, L. C. Magrini and S. U. Ahn, "Distribution Transformer Losses Evaluation: A New Analytical Methodology and Artificial Neural Network Approach," in *IEEE Transactions on Power Systems*, vol. 24, no. 2, pp. 705-712, May 2009, doi: 10.1109/TPWRS.2008.2012178.

[4] Elleuch M. and M. Poloujadoff, "Analytical model of iron losses in power transformers," in *IEEE Transactions on Magnetics*,

vol. 39, no. 2, pp. 973-980, March 2003, doi: 10.1109/TMAG. 2003.808591.

[5] Pan Chao, Kong Lingguo, LI Zhenxin, Zheng Qing, Wang Zezhong, "Analysis Based on Improved Method for Transformer Harmonic Losses," *2012 International Conference on Future Engery, Environment, and Materials, Energy Procedia* 16 (2012) 1845 – 1851.

[6] Dawood K., M. A. Cinar, B. Alboyaci and O. Sonmez, "Modelling and analysis of transformer using numerical and analytical methods," *2017 18th International Symposium on Electromagnetic Fields in Mechatronics, Electrical and Electronic Engineering (ISEF) Book of Abstracts*, Lodz, 2017, pp. 1-2, doi: 10.1109/ISEF.2017.8090696.

[7] Stoll R. L., "Approximate formula for the eddy-current loss induced in a long conductor of rectangular cross- section by a transverse magnetic field," *Proc. IEEE*, Vol. 116, No. 6, June 1969, pp. 1003- 1008.

[8] Girgis R. S., D. J. Scott, D. A. Yannucci, and J. B. Templeton, "Calculation of winding losses in shell form transformers for improved accuracy and reliability—Part I: Calculation procedure and program description," *IEEE Transactions on Power Delivery*, Vol. PWRD-2, No. 2, April 1987, pp. 398–410.

[9] Kulkarni S. V, G. S Gulwadi, R. Ramachandran, and S. Bhatia, "Accurate estimation of eddy loss in transformer windings by using FEM analysis," *International Conference on Transformers*, TRAFOTECH-94, Bangalore, India.

[10] Del Vecchio R. M, B. Poulin , Pierre T. Feghali , Dilipkumar M. Shah , Rajendra Ahuja, *"Rabins' Method for Calculating Leakage Fields, Inductances, and Forces in Iron Core Transformers, Including Air Core Methods,"* Routledge Handbooks online, August 2017.

[11] Schmidt E., P. Hamberger and W. Seitlinger, "Finite element calculation of eddy current losses in the tank wall of power

transformers," *IEEE International Electric Machines and Drives Conference*, 2003. IEMDC'03., Madison, WI, USA, 2003, pp. 1167-1173 vol.2, doi: 10.1109/IEMDC.2003.1210388.

[12] Zhao L., Q. Ge, Z. Zhou, B. Yang, K. Wang and Y. Li, "Calculation and Analysis of the Winding Loss of High-Frequency Transformer Based on Finite Element Method," *2018 21st International Conference on Electrical Machines and Systems (ICEMS)*, Jeju, 2018, pp. 2655-2658, doi: 10.23919/ICEMS.2018.8549158.

[13] Dursun K. and N. Rahmanov, "Harmonic load losses in power transformer windings using Finite Element methods," *Eurocon 2013*, Zagreb, 2013, pp. 1526-1530, doi: 10.1109/EUROCON.2013.6625180.

[14] Najafi A. and I. Iskender, "Reducing losses in distribution transformer using 2605SA1 amorphous core based on time stepping finite element method," *2015 International Siberian Conference on Control and Communications (SIBCON)*, Omsk, 2015, pp. 1-4, doi: 10.1109/SIBCON.2015.7146963.

[15] Zhao L., Q. Ge, Z. Zhou, B. Yang, K. Wang and Y. Li, "Calculation and Analysis of the Winding Loss of High-Frequency Transformer Based on Finite Element Method," *2018 21st International Conference on Electrical Machines and Systems (ICEMS)*, Jeju, 2018, pp. 2655-2658, doi: 10.23919/ICEMS.2018.8549158.

[16] Nazmunnahar M., S. Simizu, P. R. Ohodnicki, S. Bhattacharya and M. E. McHenry, "Finite-Element Analysis Modeling of High-Frequency Single-Phase Transformers Enabled by Metal Amorphous Nanocomposites and Calculation of Leakage Inductance for Different Winding Topologies," in *IEEE Transactions on Magnetics*, vol. 55, no. 7, pp. 1-11, July 2019, Art no. 8401511, doi: 10.1109/TMAG.2019.2904007.

[17] Thango B. A., J. A. Jordaan and A. F. Nnachi, "Service Life Estimation of Photovoltaic Plant Transformers Under Non-

Linear Loads," *2020 IEEE PES/IAS PowerAfrica*, Nairobi, Kenya, 2020, pp. 1-5, doi: 10.1109/PowerAfrica49420.2020.9219912.

[18] Thango B. A., J. A. Jordaan and A. F. Nnachi, "Effects of Current Harmonics on Maximum Loading Capability for Solar Power Plant Transformers," *2020 International SAUPEC/RobMech/PRASA Conference*, Cape Town, South Africa, 2020, pp. 1-5, doi: 10.1109/SAUPEC/RobMech/ PRASA48453.2020.9041101.

[19] Thango B. A., J. A. Jordaan and A. F. Nnachi, "Step-Up Transformers for PV Plants: Load Loss Estimation under Harmonic Conditions," *2020 19th International Conference on Harmonics and Quality of Power (ICHQP)*, Dubai, United Arab Emirates, 2020, pp. 1-5, doi: 10.1109/ICHQP46026.2020.9177938.

[20] Thango B. A, J. A Jordaan, A. F Nnachi, D. B Nyandeni "Solar Power Plant Transformer Loss Calculation under Harmonic Currents using Field Element Method," *9th CIGRE Southern Africa Regional Conference*, 1st – 4th October 2019, Johannesburg, South Africa.

[21] Damjanovic A., R. Integlia and A. Sarwat, "Evaluation of power transformer loses measurements methods under nonlinear load conditions," *2016 IEEE/IAS 52nd Industrial and Commercial Power Systems Technical Conference* (I&CPS), Detroit, MI, 2016, pp. 1-5, doi: 10.1109/ICPS.2016.7490230.

[22] Jean-Pierre Keradec, "Validating the power loss model of a transformer by measurement," I*EEE Industry Applications Magazine*, JULY|AUG 2007 • WWW.IEEE.ORG/IAS.

[23] Frelin W., L. Berthet, M. Petit and J. C. Vannier, "Transformer winding losses evaluation when supplying nonlinear load," *2009 44th International Universities Power Engineering Conference (UPEC)*, Glasgow, 2009, pp. 1-5.

[24] Dimitrakakis G. S., E. C. Tatakis and E. J. Rikos, "A new model for the determination of copper losses in transformer windings with arbitrary conductor distribution under high frequency sinusoidal excitation," *2007 European Conference on Power Electronics and Applications*, Aalborg, 2007, pp. 1-10, doi: 10.1109/EPE.2007.4417574.

[25] Kubota K., K. Shimura," Examination on Copper Loss Reduction of High-frequency Transformers for Trains Using Magnetic Flux Path Control Technology," *2019 International Conference on Electrical Engineering Research & Practice (ICEERP), Published 2019, Materials Science*, DOI:10.1109/ICEERP49088.2019.8956970Corpus ID: 210696152.

[26] Del Vecchio R. M., B. Poulin, P. T. Feghali, D. M. Shah, and R. Ahuja, *"Transformer design principles: with applications to core-form power transformers,"* CRC press, 2010.

[27] Zhu Z., D. Xie, G. Wang, Y. Zhang, and X. Yan, "Computation of 3-d magnetic leakage field and stray losses in large power transformer," *IEEE Trans. Magn.*, vol. 48, no. 2, pp. 739–742, Feb 2012.

[28] Li L., W. N. Fu, S. L. Ho, S. Niu, and Y. Li, "Numerical analysis and optimization of lobe-type magnetic shielding in a 334 mva single-phase auto-transformer," *IEEE Trans. Magn.*, vol. 50, no. 11, pp. 1–4, Nov 2014.

[29] Smajic J., G. D. Pino, C. Stemmler,W. Mnig, and M. Carlen, "Numerical study of the core saturation influence on the winding losses of traction transformers," *IEEE Trans. Magn.*, vol. 51, no. 3, pp. 1–4, March 2015.

[30] Bai B., Z. Chen, and D. Chen, "Dc bias elimination and integrated magnetic technology in power transformer," *IEEE Trans. Magn.*, vol. 51, no. 11, pp. 1–4, Nov 2015.

[31] Sai Siddu P. R., S. Ravi Chandran and S. Usa, "Eddy Current and Magneto-Structural Analysis on Transformer Winding with Continuously Transposed Conductors," *2019 IEEE 4th*

International Conference on Condition Assessment Techniques in Electrical Systems (CATCON), Chennai, India, 2019, pp. 1-6, doi: 10.1109/CATCON47128.2019.CN0092.

[32] Moghaddami M., A. I. Sarwat and F. de Leon, "Reduction of Stray Loss in Power Transformers Using Horizontal Magnetic Wall Shunts," in *IEEE Transactions on Magnetics*, vol. 53, no. 2, pp. 1-7, Feb. 2017, Art no. 8100607, doi: 10.1109/TMAG.2016.2611479.

[33] Djurovic M. and J. Monson, "3-dimensional computation of the effect of the horizontal magnetic shunt on transformer leakage fields," *IEEE Trans. Magn.*, vol. 13, no. 5, pp. 1137–1139, Sep 1977.

[34] Djurovic M. and J. E. Monson, "Stray losses in the step of a transformer yoke with a horizontal magnetic shunt," *IEEE Transactions on Power Apparatus and Systems*, vol. PAS-101, no. 8, pp. 2995–3000, Aug 1982.

[35] Moghaddami M., A. I. Sarwat and F. de Leon, "Reduction of Stray Loss in Power Transformers Using Horizontal Magnetic Wall Shunts," in *IEEE Transactions on Magnetics*, vol. 53, no. 2, pp. 1-7, Feb. 2017, Art no. 8100607, doi: 10.1109/TMAG.2016.2611479.

[36] Phani Kumar Mokkapaty S., J. Weiss, A. Schramm, S. Magdaleno-Adame, J. C. Olivares-Galvan and H. Schwarz, "3D Finite Element analysis of magnetic shunts and aluminum shields in clamping frames of distribution transformers," *2015 IEEE International Autumn Meeting on Power, Electronics and Computing (ROPEC)*, Ixtapa, 2015, pp. 1-6, doi: 10.1109/ROPEC.2015.7395069.

[37] Song Z. et al. "The edge effects of magnetic shunts for a transformer tank," *2011 International Conference on Electrical Machines and Systems*, Beijing, 2011, pp. 1-4, doi: 10.1109/ICEMS.2011.6074006.

[38] Rizzo M., A. Savini, J. Turowski, "Influence of flux collectors on stray losses in transformers," *IEEE Trans. on Magn.*, Vol. 36, No. 4, pp. 1915-1918, 2000.

[39] Yan L., Eerhemubayaer, S. Xin, J. Yongteng, L. Jian, "Calculation and analysis of 3-D nonlinear eddy current field and structure losses in transformer," *IEEE International conference on Electrical Machines and Systems* (ICEMS), pp. 1-4, Beijing, China, August 2011.

[40] Mokkapaty S. P. K et al. "3D Finite Element Analysis of Magnetic Shunts and Aluminium Shields in Clamping Frames of Distribution Transformers," ROPEC 2015 - Power Systems.

[41] 60076-7 IEC: 2005 "*Part 7: Loading guide for oil-immersed power transformers*," https://webstore.iec.ch/preview/info_iec60076-7%7Bed1.0%7Den_d.pdf.

Chapter 3

DESIGN CONSIDERATIONS FOR K-RATED TRANSFORMERS

The current chapter intends to examine K-rated transformers regarding harmonic indexes, particularly with respect to K-Factor, Harmonic loss factor (HLF) and Factor K for derating the transformer. The transformer considered in this study has a rating of 630kVA, 11/0.415kV, and three-phase, liquid-filled with ONAN cooling. A comparative study is carried out based on a supplied harmonic current spectrum to calculate the additional winding Eddy losses using the K-Factor and Harmonic loss factor (HLF) methods. The results yield an error of estimate of 33% and 5% for the winding Eddy losses and total losses, respectively. Further, the chapter reveals that the studied unit needs to be de-rated to 86, 6% of nominal power rating when supplying this harmonic spectrum.

1. INTRODUCTION

The amount of solar photovoltaic energy in the South African electrical grid is rapidly increasing. The target for the year 2030, set by the Integrated Resource Plan 2019 (IRP2019) [1–2] is that of 10.5GW of South Africa's energy should be derived from solar photovoltaics. To achieve this amount of generation capacity, large and remote regions with no distractions from tall buildings and trees is critical as it has a direct role in the generation profile, grid stability and economic performance of the solar photovoltaics. This increase in the grid-connected solar PV's has led to new technical and economic challenges for electrical equipment such as the step-up transformer that are unique to conventional distribution transformers [3-4].

A basic architecture of a solar photovoltaic (PV) plant and the location of the transformer in the PV system is presented in Figure 10 below.

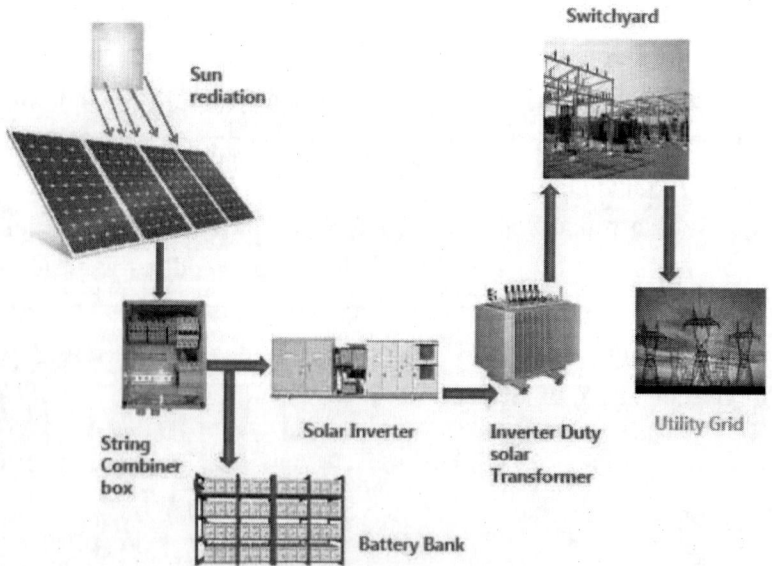

Figure 10. The basic layout of a solar photovoltaic (PV) plant [5].

The intermittent solar radiation effect makes the loading conditions of solar PV transformers distinctive from that of regular distribution transformers. The primary challenge derived by the solar radiation conditions is on the transformer no-load losses. The inverters produce only a fraction of their capacity if the level of solar radiation is under a particular value. This subject the step-up transformer to lightly loaded conditions and hence the increase in the core or no-load losses [6–8]. Additionally, the reduction of the mean load factor is directly linked to the dependency of the step-up transformer on solar radiation conditions. Unlike regular distribution transformers, the recurrent load cycle deviations take place several times during the day. The fluctuation of the load cycle culminates in the fluctuation of the current flowing within the transformer [6–8]. The load cycle generates thermal cycling and consequently the thermal ageing. The latter expedite the transformer service life by creating partial discharge, hotspot temperature rise, and degradation of the insulation [6–8]. A typical PV generation profile is shown in Figure 11 below.

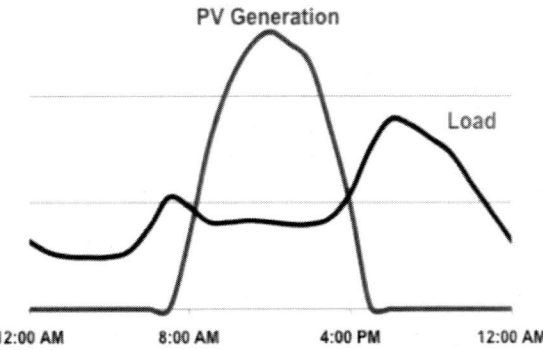

Figure 11. Solar PV generation and load cycle [9].

A harmonic profile is composed of waveforms that vary from the fundamental frequency, which is designated as 50Hz in South Africa. In solar PV plants, a fleet of inverters connected in parallel is used to deliver an a.c voltage to the step-up transformer. These inverters generate a harmonic profile of varying magnitudes at different harmonic orders that derive harmonic currents streaming through the winding conductors and other structural parts [9, 10]. Amidst earlier deployment of solar PV plants in the last decade and the growth on the use of non-linear loads distorting the sinusoidal wave within the solar PV plant technologies, the step-up transformer specifications did not entirely address the amount of harmonics that will be seen by the transformer while in service [9-10]. The resultant harmonics and distortion bring about excess temperature rise.

Additionally, if the step-up transformer is under-designed, cellulose and oil insulation will degrade and reduce the transformer service life. And so, step-up transformers purposed for solar PV technologies must be designed to meet the technical requirements of such an environment. The atypical harmonic current spectrum is shown in Figure 12 below.

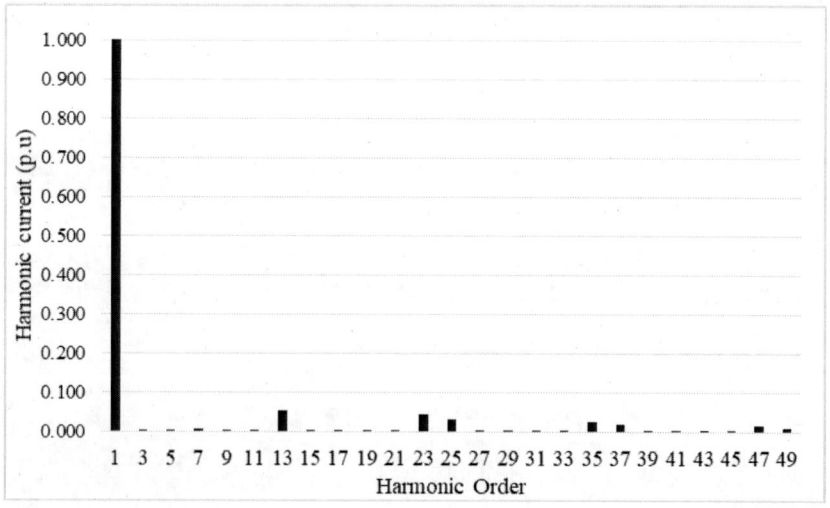

Figure 12. Harmonic current profile.

In order to design transformers capable of supplying distorted harmonic currents, procedures such as the K-Factor, harmonic loss factor and K factor methods may be employed. Authors in [11-15] have presented studies doe the evaluation of the additional losses that the transformer will see during service and methods for derating the transformer.

This chapter considers the transformer losses at fundamental frequency for a 630kVA transformer intended for solar PV application. Furthermore, the computation of the losses under the supplied harmonic profile has been conducted to investigate the service loss seen by this unit while in service. The computation of the additional losses is carried out by using the K-Factor and harmonic loss factor methods.

2. K-RATED TRANSFORMERS

K-rated transformers can handle the temperature rise effects due to the presence of harmonic load currents. Some of the considerations for these types of transformers are summarized below:

2.1. Design Considerations

A K-factor transformer design differs from that of conventional distribution transformers (CDT).

- A k-factor transformer's neutral conductor is 200 percent larger than CDT. This will assist in reducing the effects of third-order harmonics in the neutral conductor.
- The Eddy current loss can be minimized by providing electrostatic shielding between the primary and secondary windings.
- The transformer leads have been increased in size.
- A ferromagnetic material with a lower inductance is used.

- The secondary winding of the transformer has several smaller conductors to reduce the skin effect.

2.2. Harmonic Effects on the Losses

In practice, transformer manufacturers design their transformers based upon minimal losses at fundamental frequency, rated voltage and current. The presence of harmonics during service triggers an increase in the losses and subsequently the temperature rise in active metallic structures. The transformer losses may be addressed separately as the No-load loss and the load losses as shown in Eq. (1) [16].

$$P_T = P_C + P_{LL} \tag{1}$$

where,

P_T – Total losses (in kW)
P_C – No-load loss (in kW)
P_{LL} – Load loss (in kW).

It follows that the load losses can also be separated into copper losses (I^2R), winding Eddy losses (P_{EC}) and other stray losses (P_{OSL}).

In practice, the winding Eddy losses and other stray losses cannot be measured. As a result, the I^2R loss which is computed using the measured resistance and the load current is subtracted from the measured load loss to determine stray losses (winding Eddy losses and other stray losses. The loss induced by the stray electromagnetic flux in metallic structures including the winding conductors, core steel, core clamping structures, steel tank walls, et cetera, is referred to as stray losses. Eddy current due to circulating currents between strands forms part of the winding Eddy losses [17].

Design Considerations for K-Rated Transformers 43

A summary of the losses affected by the harmonic load current for transformers intended for solar PV application at the point of common coupling is as follows:

- The impact of harmonic current on I^2R loss. The I^2R loss will increase if the RMS value of the load current is increased due to harmonic component.
- Impact of harmonic current on P_{EC}. In the power frequency range, winding eddy current loss (P_{EC}) is proportional to the square of the load current and the square of frequency. This property can result in excessive winding loss and, as a result, abnormal winding temperature rise in transformers that supply load currents.
- The effect of a harmonic current on P_{OSL}. Other stray loss (P_{OSL}) in the core, clamps and structural parts is expected to increase at a rate proportional to the load current squared. However, unlike winding eddy losses, these losses will not increase at a rate proportional to the square of the frequency Eddy current loss in bus bars, linking and structural parts increases by a harmonic exponent factor of 0.8 or less studies conducted by manufacturers and other researchers.
- DC Components of load current. The presence of a dc component in the load current is normal in harmonic load currents. The transformer core loss will be marginally increased by a dc portion of load current, but the magnetizing current and audible sound level will be significantly increased. The load carrying capacity of a transformer calculated by this suggested procedure is assumed to be unaffected by relatively small dc components (up to the r.m.s magnitude of the transformer excitation current at rated voltage). Higher dc current components may have a negative impact on transformer performance and should be circumvented.

3.1. Transformers under Harmonic Conditions

3.1.1. K-Factor

The K-factor calculation for transformers intended to be in-service to solar PV plants where the transformer will be susceptible to harmonic conditions is described by the European standard [1] as shown in Eq. (2) below [17].

$$K = \sum_{h=1}^{h=h_{max}} h^2 I_h^{\ 2} \tag{2}$$

where,

h − Harmonic order
P_C − Fraction of the load current at harmonic order, h.

Depending on the value calculated based on the supplied harmonic spectrum, the K-Factor standard range of 4, 9, 13, 20, 30, 40, and 50 is described in the standard.

3.1.2. American Standards IEEE - ANSI

In the IEEE Std. C57.110-2018 [17] standard, the additional losses caused by the harmonic current content is described by the harmonic loss factor (F_{HL}), which is calculated by multiplying the current harmonics by the factor as expressed in Eq. (3) below.

$$F_{HL} = \frac{\sum_{h=1}^{h=h_{max}} (\frac{I_h}{I_1})^2 h^2}{\sum_{h=1}^{h=h_{max}} (\frac{I_h}{I_1})^2} \tag{3}$$

where,

I_h − Load current under harmonic conditions
I_1 − Load current at fundamental frequency

Design Considerations for K-Rated Transformers

The additional winding Eddy losses can be determined as follows in Eq. (4) and by multiplying the harmonic loss factor.

$$P_{EC} = P_{EC-R} \sum_{h=1}^{h=h_{max}} \left(\frac{I_h}{I_1}\right)^2 h^2 \qquad (4)$$

Similarly, the other stray losses can be ascertained as follows in Eq. (5).

$$P_{OSL} = P_{OSL-R} \sum_{h=1}^{h=h_{max}} \left(\frac{I_h}{I_1}\right)^{0.8} h^2 \qquad (5)$$

3.1.3. Factor-K

The method for derating a transformer supplying harmonic current profile is also described by the European standard [17] as the inverse of the factor K as shown in Eq. (6) and Eq. (7) [17].

$$K = \left[1 + \frac{e}{1+e}\left(\frac{I_1}{I}\right)^2 \sum_{h=2}^{h=h_{max}} h^q \left(\frac{I_h}{I_1}\right)^2\right]^{\frac{1}{2}} \qquad (6)$$

$$DR_k = \frac{1}{K} \qquad (7)$$

e – Ratio of Winding Eddy current loss to I2R loss at rated frequency and reference temperature.
I = r.m.s of the sinusoidal current including all harmonics.
q = a constant with a value of 1.7 for rectangular winding conductors.

3.2. Case Study and Results

In the current chapter, a 630kVA transformer with the rated losses specified in Table 7 is studied. The unit has a secondary full-load current of 876.485A and a turns ratio of 26.506.

Table 7. Transformer rated losses

Type of loss	Rated losses (W)
No load	689
I²R	3 826
Winding eddy	144
Other stray	192
Total losses	4 850

The copper losses are observed to have the largest contribution of 79%, followed by the no-load losses of 14%, other stray losses of 4% and the winding Eddy losses of 3%.

3.2.1. Harmonic Content

Figure 13 shows the harmonic current load spectrum that the transformer will see during service.

Figure 13. Harmonic current content.

Based on this harmonic spectrum, the p.u r.m.s current of 1.154 is calculated. The k-Factor and FHL are 7.186 and 9.577, respectively. Dividing these values by the p.u r.m.s current, the factors to calculate the additional losses will be 6.357 and 7.19, respectively.

3.2.2. Calculation of Factors

In Table 8, a comprehensive tabulation of the K-factor calculations using Eq. (2) and the supplied harmonic spectrum is presented.

Table 8. K-factor calculation

h	I_h/I_1	$(I_h/I_1)^2$	I_h/I_r	$(I_h/I_r)^2$	$(I_h/I_r)^2 h^2$
1	1	1,000	0,866	0,750	0,750
3	0,45	0,203	0,390	0,152	1,368
5	0,27	0,073	0,234	0,055	1,368
7	0,19	0,036	0,165	0,027	1,327
9	0,092	0,008	0,080	0,006	0,514
11	0,071	0,005	0,062	0,004	0,458
13	0,051	0,003	0,044	0,002	0,330
15	0,043	0,002	0,037	0,001	0,312
17	0,04	0,002	0,035	0,001	0,347
19	0,039	0,002	0,034	0,001	0,412
	$\Sigma(I_h/I_1)^2$	1,333			7,186
	$\sqrt{\Sigma(I_h/I_1)}$	1,154		K-Factor	7,186

The maximum permissible per-unit harmonic load current with the given harmonic spectrum using the K-Factor method will be 759,274A. Additionally, using the K-Factor standard range, this unit can be classified as a K-9 rating transformer.

The calculation of the harmonic loss factor (using Eq. 3) has also been demonstrated as tabulated in Table 9 below.

Table 9. Harmonic loss factor calculation

h	I_h/I_1	$(I_h/I_1)^2$	$(I_h/I_1)^2 h^2$	$(I_h/I_1)^2 h^{0.8}$
1	1	1,000	1,000	1,000
3	0,45	0,203	1,823	0,488
5	0,27	0,073	1,823	0,264
7	0,19	0,036	1,769	0,171
9	0,092	0,008	0,686	0,049
11	0,071	0,005	0,610	0,034
13	0,051	0,003	0,440	0,020
15	0,043	0,002	0,416	0,016
17	0,04	0,002	0,462	0,015
19	0,039	0,002	0,549	0,016
	$\Sigma(I_h/I_1)^2$	1,333	9,577	2,074
	$\sqrt{\Sigma(I_h/I_1)}$	1,154		

The maximum permissible per-unit harmonic load current with the given harmonic spectrum using the Harmonic loss factor method will be 652A.

3.2.3. Service Losses

The additional losses seen by the transformer during service are calculated using the K-Factor and Harmonic loss factor method as tabulated in Table 10 below.

The results yield an error of estimate of 33% and 5 % for the winding Eddy losses and total losses, respectively.

A comparison of the methods considered in this study is tabulated as shown in Table 11.

Table 10. Transformer rated losses

Type of loss	Rated losses (W)	Load losses (W)	Corrected losses (W)
No load	689	689	689
I²R	3 826	5 098	5 098
Winding eddy (K-Factor)	144	191	1 374
Winding eddy (HLF)	144	191	1 031
Other stray	192	256	398
Total losses (K-Factor)	4 850	6 234	7 216
Total losses (HLF)	4 850	6 234	7 559

Table 11. Transformer rated losses

Factor	Value	Rated current (I_r) (A)
K-Factor	5,393	759
HLF	7,19	652

An Error estimate of 16% is observed between the compared methods in respect to the rated currents.

3.2.4. Derating

The procedure for the amount of derating required when supplying the harmonic spectrum in Figure 4 is detailed in Table 12 below using Eq. (6) and Eq. (7).

In practice, the studied transformer would need to be de-rated to (1/1.15) 86, 6% of nominal power rating when supplying this spectrum. The latter yield a de-rated power rating of 546kVA.

Table 12. Factor-k calculation

h	I_h/I_1	$(I_h/I_1)^2$	h^q	$(I_h/I_1)^2 h^q$
1	1	1,000	1,000	1,000
3	0,45	0,203	6,473	1,311
5	0,27	0,073	15,426	1,125
7	0,19	0,036	27,332	0,987
9	0,092	0,008	41,900	0,355
11	0,071	0,005	58,934	0,297
13	0,051	0,003	78,290	0,204
15	0,043	0,002	99,852	0,185
17	0,04	0,002	123,527	0,198
19	0,039	0,002	149,239	0,227
	$\Sigma(I_h/I_1)^2$	1,333		
	$\sqrt{\Sigma(I_h/I_1)}$	1,154	[a]	4,887
	Ratio (Above)	0,750	[a]* Ratio	3,667
			e/(e+1)	0,091
			K^2	1,333
			K	1,15

CONCLUSION

This chapter presents the computation of the losses for a 630kVA transformer using the K-factor and Harmonic loss factors methods. Based on a supplied harmonic current spectrum, a comparative study is carried out to calculate the additional winding Eddy losses using the K-Factor and Harmonic loss factor (HLF) methods. The results yield an error of estimate of 33% and 5% for the winding Eddy losses and total losses, respectively. Further, the chapter reveals that the studied unit needs to be de-rated to 86, 6% of nominal power rating when supplying this harmonic spectrum.

REFERENCES

[1] Akom K., T. Shongwe, M. K Joseph "South Africa's integrated energy planning framework, 2015–2050" *Journal of Energy in Southern Africa*, Vo. 32, issue 1, Feb., 2021. [Online]. Available: www.iisd.org.

[2] van der Poel R. J., A. Felekis, "What you need to know: South Africa's Integrated Resource Plan 2019," in *Mining review Africa*, Oct. 2019. [Online]. Available: www.miningreview.com.

[3] Thango B. A., J. A. Jordaan and A. F. nnachi, "Contemplation of Harmonic Currents Loading on Large-Scale Photovoltaic Transformers," *2020 6th IEEE International Energy Conference (ENERGYCon)*, Gammarth, Tunisia, 2020, pp. 479-483, doi: 10.1109/ENERGYCon48941.2020.9236514.

[4] Che Wanik M. Z., M. M. Bukshaisha and S. R. Chaudhry, "PV generation in distribution network and its impact on power transformer on-load tap changer operation," *2017 IEEE Manchester PowerTech*, Manchester, UK, 2017, pp. 1-6, doi: 10.1109/PTC.2017.7981210.

[5] Agrawal Y., K. S. Kumar, "Ability To Withstand Short-Circuit Test On Inverter-Duty Transformer For Solar Application," in *Electrical India*, Sept., 2019. [Online]. Available: www.electricalindia.in.

[6] Thango B. A., J. A. Jordaan and A. F. Nnachi, "Effects of Current Harmonics on Maximum Loading Capability for Solar Power Plant Transformers," *2020 International SAUPEC/RobMech/PRASA Conference*, Cape Town, South Africa, 2020, pp. 1-5, doi: 10.1109/SAUPEC/RobMech/ PRASA48453.2020.9041101.

[7] Deokar S. A. and L. M. Waghmare, "Impact of power system harmonics on insulation failure of distribution transformer and its remedial measures," *2011 3rd International Conference on*

Electronics Computer Technology, Kanyakumari, India, 2011, pp. 136-140, doi: 10.1109/ICECTECH.2011.5941817.

[8] Sumaryadi, H. Gumilang and A. Suslilo, "Effect of power system harmonic on degradation process of transformer insulation system," *2009 IEEE 9th International Conference on the Properties and Applications of Dielectric Materials*, Harbin, China, 2009, pp. 261-264, doi: 10.1109/ICPADM.2009.5252458.

[9] Firat Y., "Utility-scale solar photovoltaic hybrid system and performance analysis for eco-friendly electric vehicle charging and sustainable home," *Energy Sources Part A Recovery Utilization and Environmental Effects*, Vol. 41, Issue 4, Page 1-12. doi: 10.1080/15567036.2018.1520354.

[10] Thango B. A., K. Moloi, J. A. Jordaan and A. F. Nnnach, "A Further Look into the Service Lifetime Cost of Solar Photovoltaic Energy Transformers," *2021 Southern African Universities Power Engineering Conference/Robotics and Mechatronics/Pattern Recognition Association of South Africa (SAUPEC/RobMech/PRASA)*, Potchefstroom, South Africa, 2021, pp. 1-7, doi: 10.1109/SAUPEC/RobMech/PRASA52254.2021.9377229.

[11] Yildirim D. and E. F. Fuchs, "Measured transformer derating and comparison with harmonic loss factor (F/sub HL/) approach," in *IEEE Transactions on Power Delivery*, vol. 15, no. 1, pp. 186-191, Jan. 2000, doi: 10.1109/61.847249.

[12] Fuchs E. F., Dingsheng Lin and J. Martynaitis, "Measurement of three-phase transformer derating and reactive power demand under nonlinear loading conditions," in *IEEE Transactions on Power Delivery*, vol. 21, no. 2, pp. 665-672, April 2006, doi: 10.1109/TPWRD.2005.858744.

[13] Cherian E. and G. R. Bindu, "Minimizing Harmonics and Transformer Derating in Low Voltage Distribution Networks by DC Distribution," *2018 International Conference on Emerging*

Trends and Innovations In Engineering And Technological Research (ICETIETR), Ernakulam, India, 2018, pp. 1-6, doi: 10.1109/ICETIETR.2018.8529063.

[14] Tao Shun and Xiao Xiangning, "Comparing transformer derating computed using the harmonic loss factor FHL and K-factor," *2008 Third International Conference on Electric Utility Deregulation and Restructuring and Power Technologies*, Nanjing, China, 2008, pp. 1631-1634, doi: 10.1109/DRPT.2008.4523666.

[15] Cazacu E. and L. Petrescu, "Derating the three-phase power distribution transformers under nonsinusoidal operating conditions: A case study," *2014 16th International Conference on Harmonics and Quality of Power (ICHQP)*, Bucharest, Romania, 2014, pp. 488-492, doi: 10.1109/ICHQP.2014.6842930.

[16] Thango B. A., J. A. Jordaan and A. F. Nnachi, "Selection and Rating of the Step-up Transformer for Renewable Energy Application," in *SAIEE Africa Research Journal*, vol. 111, no. 2, pp. 50-55, June 2020, doi: 10.23919/SAIEE.2020.9099492.

[17] "IEEE Recommended Practice for Establishing Liquid-Immersed and Dry-Type Power and Distribution Transformer Capability When Supplying Nonsinusoidal Load Currents," in *IEEE Std C57.110™-2018* (Revision of IEEE Std C57.110-2008), vol., no., pp. 1-68, 31 Oct. 2018, doi: 10.1109/IEEESTD.2018.8511103.

Chapter 4

TRANSFORMER THERMAL PERFORMANCE

Recently, there's been a rapid increase in demand for Distributed Photovoltaic Power (DPVP) generation system transformers and the rise in the construction of solar photovoltaic plants in South Africa. However, there've been reported technical challenges of DPVP generation system transformers with high top-liquid and hotspot temperatures during their service life. The winding Eddy current loss harmonic factors recommended by the standards do not take into account the skin effect due to the restricted penetration of electromagnetic fields on the conductors at high harmonic orders, which results in the erroneous estimation of these losses as well as high-temperature rise. In order to enable optimum estimation of the winding Eddy current losses, it is proposed in this paper a harmonic loss factor that considers the conductor skin effect under harmonic conditions by using the data supplied by the manufacturer for losses under rated conditions. Based on supplied harmonic spectrum, the service losses of a liquid-filled step-up transformer are estimated, top-liquid and hotspot temperature conditions are then realized.

Lastly, the capacity of a transformer when supplying the given distorted load current profile is evaluated by using a technique ascribed to derating. The use of this method is achieved by the computation of the maximum permissible current as a result of increased winding Eddy current loss and other structural losses. The results indicate transformer loading capacity increase to 97.6% compared to 91.3% when applying the harmonic factor approach recommended by the standards.

1. INTRODUCTION

Presently, South Africa supplements more renewable energy capacity annually than new additional capacity from coal energy generation (CEG) [1, 2]. Because South Africa is predominantly dependent on CEG which amounts to around 90 % of its generation capacity, about 45 % of the country's carbon emissions are also due to CEG [3]. A paradigm shift to ascertain future energy security by overthrowing CEG with renewable energy technologies is considered the most efficient means to eradicate the greenhouse effect [4].

South Africa is situated in a favourable location that possesses affluence of renewable energy resources, especially wind and solar. In essence, reducing CEG and expanding the renewable energy sources (RES) is a 'low-hanging fruit ready for harvest. This is underpinned by the successful implementation of the Integrated Resource Plan (IRP) [4] and the Renewable Energy Independent Power Producer Procurement Programme (REIPPPP) [5], through which energy security, eradication of green effect and socio-economic growth are prioritized.

The Independent Power Producers (IPP's) have been the driving force behind the commissioning of large-scale solar and wind generation plants [6]. Most of the metropolitan municipalities are even exploring alternatives to purchase power from the IPP's. The evolvement of renewable energy market does, however, present some technical challenges to the IPP's especially on the electrical equipment operated within solar PV energy generation plants.

The Distributed Photovoltaic Power (DPVP) transformer is an important component in a solar photovoltaic plant. DPVP transformers that produce distorted load currents are increasing in solar photovoltaic plants [7-9]. The top-liquid and hotspot temperature conditions are highly strained with the effect of the harmonic profiles. The solar electromagnetic irradiation is not steady all throughout the day, due to which the output of the solar module varies. Various solar PV plants in South Africa have been identified to have DPVP transformers that are failing at an alarming rate [10, 11].

The major reason for these failures is linked to the general practice among IPP's to use regular distribution transformers in the solar PV plants. In essence regular distribution transformers are not capable to endure the adverse effects of variable inverters and non-linear load outputs [7]. The DPVP transformers are challenged with problems of thermal ageing of the insulation material due to the occurrence of high winding losses and other metal parts losses. The electrical problems may be intermittent loading cycle, harmonics distortion and voltage transients. These narrow requirements are not often discerned by transformer manufacturers when the information is not included in the technical specification supplied by the utility owners at tender stage. As well, if the utility owners did not perform a complete harmonic study of the plant, the reliability of the step-up transformer tends to be lower than those used for non-renewable energy applications.

In this chapter, a harmonic spectrum supplied by the IPP is considered to estimate the service losses of a liquid-filled step-up transformer. From the losses, the top-liquid and hotspot temperature conditions are then realized. Lastly, an evaluation of the capacity of a transformer when supplying a given distorted load current profile is undertaken by using a technique ascribed to as derating.

2. TRANSFORMER LOSSES AND HOTSPOT TEMPERATURE

Transformer total loss can be described as the arithmetic sum of the no-load loss (P_{NL}) and load loss (P_{LL}). The load loss is subdivided into winding copper loss (P_{cu}) and stray loss. Additionally, the stray loss is further subdivided into the winding Eddy-current loss (P_{WEC}) and other structural parts loss (P_{OSL}). The stray loss is the energy loss as a result of electromagnetic flux impinging upon the surface of winding conductors, core laminations, flitch plates, core clamps, and tank walls. The transformer total loss can then be expressed as follows in eq. (1) [6, 7]:

$$P_{Total} = P_{NL} + P_{LL} = P_{NL} + P_{cu} + P_{WEC} + P_{OSL} \qquad (1)$$

In practice, the copper loss is provided by the transformer manufacturer after performing the load loss test or they can be calculated using the rated r.m.s current (I_R) and winding resistance (R_{DC}), which can be obtained during the resistance test. This resistance varies with harmonic order (h) as a result of the skin and proximity effect. The copper loss can be expressed as follows in eq. (2) [12, 13]:

$$P_{cu} = R_{DC} \times \sum_{h=1}^{h=h_{max}} I_{h,max}^2 \qquad (2)$$

where, I_h is the harmonic current component that will be seen by the transformer in-service.

The winding Eddy current loss under harmonic condition has the tendency to increase with the load current and nearly progressive to the square of the harmonic order as described in eq. (3), where P_{WEC_R} is rated P_{WEC}. This consequence in the spread of temperature rise upon the winding conductors and hotspot temperature [12, 13].

$$P_{WEC} = P_{WEC\_R} \times h^2 \left[\frac{I_h}{I_R}\right]^2 \tag{3}$$

The winding Eddy current loss factor, which accounts for the additional losses as a result of harmonic current generated by solar PV environment is expressed as follows in eq. (4) [12, 13]:

$$F_{WEC} = \frac{\sum_{h=1}^{h=h_{max}} h^2 \left[\frac{I_h}{I_R}\right]^2}{\sum_{h=1}^{h=h_{max}} \left[\frac{I_h}{I_1}\right]^2} \tag{4}$$

The main limitation of eq. (3) is that it is the assumption of the harmonic factor to progress by a square of the harmonic order is only suitable for conductors of up to 3 millimeters (mm). In the case of large winding conductors, this formulation leads to conservative assumptions and significant errors. The latter can lead to overdesigning of the transformer which is not favourable for transformer manufactures competing for a bid. The IPP use the losses as part of the selection criterion, and the low loss transformers means low operation cost from their perspective. In order to close this knowledge gap for large conductors, a function is adopted from [14]. In this chapter, this function is treated as an improved harmonic factor as follows in eq. (5).

$$F^*_{WEC} = \frac{\sum_{h=1}^{h=h_{max}} \left(\frac{F(\varphi_h)}{F(\varphi_R)}\right)\left(\frac{I_h}{I_1}\right)^2}{\sum_{h=1}^{h=h_{max}} \left(\frac{I_h}{I_1}\right)^2} \tag{5}$$

where,

$$F(\varphi) = \frac{1}{\varphi}\frac{\sinh\varphi - \sin\varphi}{\cosh\varphi + \cos\varphi} \tag{6}$$

and,

$$\varphi = \frac{H_c}{\delta} \tag{7}$$

where, H_c is the field strength and δ is depth of penetration.

At rated conditions, the depth of penetration can then be expressed as follows in Eq. (8).

$$\delta_R = \sqrt{\frac{\rho}{\pi \mu_o f}} \tag{8}$$

where, ρ is resistivity, μ_o is permeability of free space and f is the fundamental frequency.

The depth of penetration under harmonic conditions can be obtained as follows:

$$\delta_h = \frac{H_c \times \sqrt{h}}{\delta_R} = \delta_R \times \sqrt{h} \tag{9}$$

The improved winding eddy current losses can then be expressed as follows:

$$P^*{}_{WEC} = P_{WEC\_R} \times F^*{}_{WEC} \tag{10}$$

The eddy currents also penetrate other structural parts in the active part including the tank walls, core clamps, flitch plates, etc. The computation of other structural losses to account for harmonic condition is expressed as follows:

$$F_{OSL} = \frac{\sum_{h=1}^{h=h_{max}} h^{0.8} \left[\frac{I_h}{I_1}\right]^2}{\sum_{h=1}^{h=h_{max}} \left[\frac{I_h}{I_1}\right]^2} \tag{11}$$

Other structural parts losses can be expressed as the product of the rated structural part losses and the harmonic factor as follows in eq. (12).

$$P_{OSL} = P_{OSL\_R} \times F_{OSL} \tag{12}$$

It is also well established that the above-mentioned harmonic losses culminate in a significant temperature rise in the active part during transformer service life. In [15], the IEC 60076-7 presented a model based in Figure 14 to approximate the transformer temperature rise under harmonic conditions. This model adopts the calculation procedure of the hotspot temperature rise as described in [16]. The theoretical development of the thermal model considers the circulation of oil and temperatures location enclosed within the transformer tank.

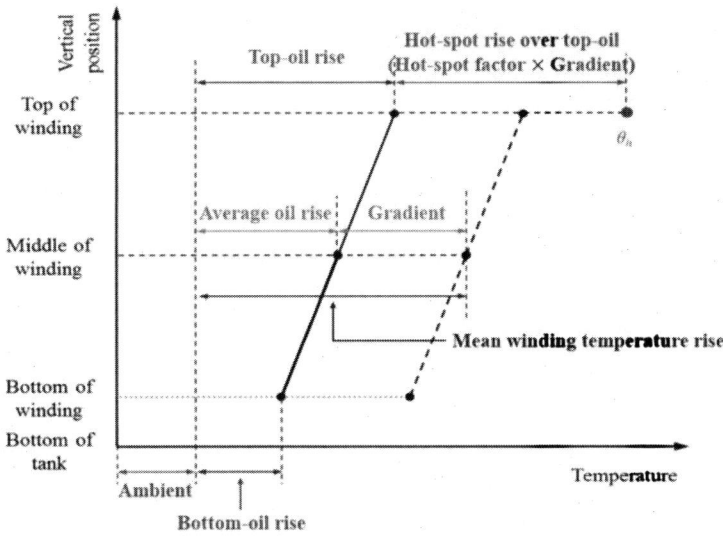

Figure 14. Transformer thermal model [15].

The effect of the transformer losses on the top-liquid (oil) temperature rise (θ_{TO}) can then be expressed as follows in eq. (13) [12, 13]. This temperature depends upon the transformer specific design parameters. Under rated conditions, the top-liquid temperature (θ_{TO_R}) is defined as the hottest-spot conductor rise over top-liquid temperature.

$$\theta_{TO} = \theta_{TO\_R} \times \left[\frac{P_{LL}+P_{NL}}{P_{LL\_R}+P_{NL}}\right]^{0.8} \quad (13)$$

The temperature gradient under rated conditions is defined as the top-liquid-rise over ambient temperature as follows in eq. (14) [12], [13].

$$\theta_g = \theta_{g\_R} \times \left[\frac{P_{LL}}{P_{LL\_R}}\right]^{0.8} \quad (14)$$

The hot spot temperature is expressed as follows in eq. (15) below. In practice, the hotspot temperature is obtained by measuring the winding resistance during the heat run test measurements.

$$\theta_{HS} = \theta_A + \theta_{TO} + \theta_g \quad (15)$$

3. CASE STUDY: 2500 KVA DPVP TRANSFORMER

The harmonic spectrum of the solar PV plant shown in Figure 16, located in Northern Cape Province, South Africa, is shown in Figure 15 to calculate the service loss, thermal performance, and capability of a liquid-filled transformer with a rated current of 131.22 A on the secondary winding.

Figure 3 shows the behavioural switching action patterns of the non-linear loads that are connected in the solar PV plant.

Figure 15. Solar PV plant.

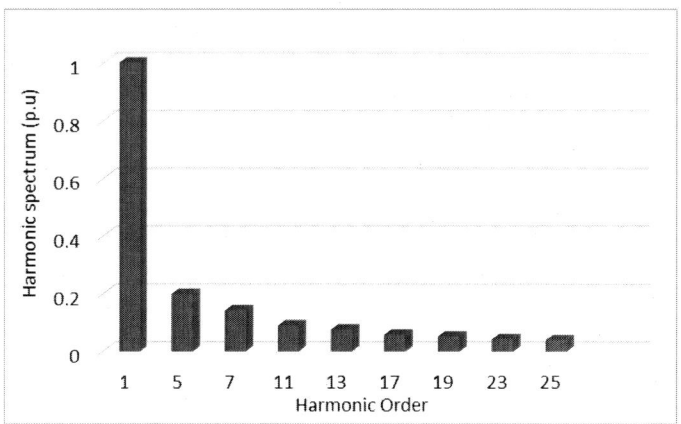

Figure 16. Harmonic spectrum of 2500 kVA DPVP transformer.

The rated, load and service loss distribution of the transformer under study is demonstrated in Table 13. The load loss was attained at 75 % load capacity and standardized to the r.m.s current. The per-unit (p.u.) r.m.s current due to the supplied harmonic spectrum is 1.041 p.u. The winding Eddy current loss and structural part loss harmonic factors are 9 and 1.415, respectively. Dividing these factors by the peak load current, 1.084 p.u., result in the service loss factors of 8.3 and 1.31, respectively as tabulated below.

Table 13. Tabulated Loss Computation under Normal Harmonic Spectrum

Type of loss	Rated losses (W)	Load losses (W)	Harmonic factor	Service losses (W)
No-load	3 981	3 981	-	3 981
I^2R	14 342	15 551	-	15 551
Winding Eddy	385	417.68	8.3	3 466
Other stray	1 064	1 154	1.31	1 544
Total losses	19 772	17 123	-	24 543

The conductor height of the low voltage winding is 11.3 mm, and the depth of penetration at fundamental frequency by using eq. (8) is 10.63 mm. Moreover, the improved harmonic factor of the winding eddy losses is 2.274. Dividing this factor by the load current (1.084 p.u.) the improved service loss factor is 2.097.

Table 14. Tabulated Loss Computation under Improved Harmonic Spectrum

Type of loss	Rated losses (W)	Load losses (W)	Harmonic factor	Service losses (W)
No-load	3 981	3 981	-	3 981
I^2R	14 342	15 551	-	15 551
Winding Eddy	385	417.68	2.097	876
Other stray	1 064	1 154	1.34	1 544
Total losses	19 772	17 123	-	21 952

The resultant winding Eddy losses due to the improved harmonic factor in Table 14 yield a 74.73% improvement of the winding Eddy losses compared to Table 13. Additionally, a reduction of 10.56% can be observed on the transformer total losses between Table 13 and Table 14.

4. DPVP Transformer Thermal Performance

This section details the thermal requirement of the transformer under study after employing the harmonic spectrum in Table 13. Under rated conditions, the transformer has a top liquid rise of 42.1°C and hotspot temperature is 51.5°C. The resultant top liquid temperature rise under harmonic condition is computed as follows:

$$\theta_{TO} = \theta_{TO\_R} \times \left[\frac{P_{LL}+P_{NL}}{P_{LL\_R}+P_{NL}}\right]^{0.8} = \theta_{TO\_R} \times \left[\frac{P_{LL}+P_{NL}}{P_{LL\_R}+P_{NL}}\right]^{0.8} = 50.05°C \quad (16)$$

$$\theta_g = \theta_{g\_R} \times \left[\frac{P_{LL}}{P_{LL\_R}}\right]^{0.8} = 10.81°C \quad (17)$$

The Hot-spot temperature for using the losses in Table 14 is computed as follows:

$$\theta_{HS} = \theta_A + \theta_{TO} + \theta_g = 60.86°C \quad (18)$$

The top-liquid temperature that will be experienced by the transformer at the top layer of the oil and tank while in service using the improved harmonic factor to account for Eddy current losses during the tender stage is as follows:

$$\theta_{TO} = \theta_{TO\_R} \times \left[\frac{P_{LL}+P_{NL}}{P_{LL\_R}+P_{NL}}\right]^{0.8} = 42.1 \times \left[\frac{P_{LL}+P_{NL}}{P_{LL\_R}+P_{NL}}\right]^{0.8} = 45.77 \quad (19)$$

$$\theta_g = 9.4 \times \left[\frac{P_{LL}}{P_{LL\_R}}\right]^{0.8} = 9.62 \text{ °C} \quad (20)$$

The hot-spot temperature of the winding conductor, which is in direct contact with the transformer oil using the losses in Table 14 are computed as follows:

$$\theta_{HS} = \theta_{TO} + \theta_g = 55.39\ °C \tag{21}$$

The service's losses due to the improved harmonic factor in Table 14 reveal that the temperature is improved by nearly 8.99 % compared to the losses estimated using the h^2 approximation in Table 13. It follows that a more suitable cooling system can be designed for the transformer under study to avoid the deterioration of the cellulosic and liquid insulation.

5. DPVP Transformer Derating

The distorted harmonic current spectrum does not only bring about the rise in temperature of DPVP transformers. The distorted current also bring about changes in the rated continuous power rating capacity and the maximum permissible current of the solar PV transformer. This phenomenon is described as transformer derating and is established on the computation of the maximum permissible per-unit current as a result of increased per-unit winding Eddy current loss ($P_{WEC\_R}(p.u)$) and loss in other structural parts ($P_{OSL\_R}(p.u)$) [12, 13]. In order to evaluate the capacity of a transformer when supplying a distorted load current, the total load current is expressed as follows in eq. (22).

$$P_{LL\_R(p.u)} = \left(1 + P_{WEC\_R}(p.u) + P_{OSL\_R}(p.u)\right) \tag{22}$$

In this case, the copper losses are defined as 1 on account of linear loads. Subsequently, the winding Eddy loss and Other Structural parts losses in eq. (20) are substituted with the correct combination of (3), (4), (11) and (12) in per-unit [12, 13]. The winding Eddy loss in per unit is then expressed as follows in eq. (23).

$$P_{EC(p.u)} = I^2_{rms(p.u)} \times \left(1 + F_{WEC} \times P_{WEC\_R}(p.u) + F_{OSL} \times P_{OSL\_R}(p.u)\right) \tag{23}$$

Eq. (21) then enables the formulation of the calculation procedure for the maximum permissible per-unit distorted load current as expressed in eq. (24) [12, 13].

$$I_{max}(pu) = \sqrt{\frac{P_{LL-R}(p.u)}{1 + F_{WEC} \times P_{EC\_R}(p.u) + F_{OSL} \times P_{OSL\_R}(p.u)}} \tag{24}$$

The corresponding maximum amperage current is expressed as follows in eq. (25).

$$I_{max} = I_{max(p.u)} \times I_{rated} \tag{25}$$

In order to calculate the derating for the liquid-filled transformer under study due to the harmonic profile in Figure 2, eq. (22) is employed using the rated transformer losses of Table 13, which yields

$$P_{LL\_R(p.u)} = 1 + 0.027 + 0.074 = 1.101 \; p.u$$

The winding Eddy current loss in per-unit using eq. (23) is given as

$$P_{EC(p.u)} = 1 + 8.3 \times 0.027 + 1.34 \times 0.074 = 1.323 \; p.u$$

The maximum permissible per-unit distorted load current using eq. (24) is computed as follows:

$$I_{max}(pu) = \sqrt{\frac{1.101}{1.324}} = 0.913 \, p.u$$

$$I_{max} = 0.913 \times 131.22 = 119.75 \, A$$

To such an extent, the result points out that the maximum permissible current is reduced to 119.75 A and the transformer capacity is now nearly 91.3 % of its rated load current capability. The same procedure for transformer losses using the improved harmonic factor in Table II is applied.

$$P_{LL\_R(p.u)} = 1 + 0.027 + 0.074 = 1.101 \, p.u.$$

The winding Eddy loss in per-unit using eq. (21) is obtained as follows:

$$P_{EC(p.u)} = 1 + 2.097 \times 0.027 + 1.34 \times 0.074 = 1.156 \, p.u$$

$$I_{max}(pu) = \sqrt{\frac{1.101}{1.156}} = 0.976 \, p.u$$

$$I_{max} = 0.976 \times 131.22 = 128 \, A$$

For the transformer losses using the improved harmonic factor, this result reveal that the maximum permissible current is increased to 128 A and the transformer loading capacity is now about 97.6% of its rated load current capability.

CONCLUSION

This chapter demonstrates the advantage of using an improved harmonic loss factor to calculate the service losses, hotspot temperature rise, and the capability of DPVP transformers with inherent distorted load current. With the improved harmonic factor, it is possible to accurately estimate the service losses for DPVP transformer winding conductors with large dimensions by taking into account the skin and proximity effect of the magnetic flux at high harmonics. A reduction of about 10.56% was observed on the transformer total losses between the normal (h^2) and improved harmonic factor approximation methods. The improved harmonic loss factor has higher accuracy than the h2 approximation because it helps to foresee any hotspot temperature rise and, consequently, avoid the transformer insulation deterioration. The latter approximation method also allows conclusions about the transformer capability when supplying distorted harmonic current. Between h^2 and improved harmonic factor approximation methods, increments of about 13.62% and 6.51% for the winding Eddy current loss and the maximum permissible currents, respectively, were obtained. It is further noted that the rated current against the maximum current is de-rated by 8.74% and 2.39% for the h^2 and improved harmonic factor approximation methods, respectively. Overall, the chapter shows an increased prediction accuracy necessary for the design reliability of the emerging large distribution transformer application areas.

REFERENCES

[1] Thopil G. A. and A. Pouris, "Externality valuation in South Africa's coal based electricity generation sector," *Proceedings of*

PICMET '11: Technology Management in the Energy Smart World (PICMET), Portland, OR, USA, 2011, pp. 1-6.

[2] Department of Energy, "2018 South African Energy Sector Report," *Energy Data Collection, Management and Analysis*, Available online: http://www.energy.gov.za.

[3] Ireland G. and J. Burton, *An assessment of new coal plants in South Africa's electricity future*, Energy Research Centre University of Cape Town, May 2018.

[4] IRP-2019: *INTEGRATED RESOURCE PLAN – Government Gazette, Department of Energy, Republic of South Africa*, Vol. 652, No. 42778, 18 October 2019. Last accessed online on 2020/04/26 at: http://www.energy.gov.za/IRP/2019/IRP-2019.pdf.

[5] Department of Energy, *Independent Power Producers Procurement Programme (IPPPP) – An overview*, March, 2017.

[6] Buraimoh E., A. A. Adebiyi, O. J. Ayamolowo and I. E. Davidson, "South Africa Electricity Supply System: The Past, Present and The Future," *IEEE PES/IAS PowerAfrica*, Nairobi, Kenya, 2020, pp. 1-5, doi: 10.1109/PowerAfrica49420.2020.9219923.

[7] Sikhosana L. S., U. B. Akuru and B. A. Thango, "Analysis and Harmonic Mitigation in Distribution Transformers for Renewable Energy Applications," *Southern African Universities Power Engineering Conference/Robotics and Mechatronics/ Pattern Recognition Association of South Africa (SAUPEC/ RobMech/PRASA)*, Potchefstroom, South Africa, 2021, pp. 1-6.

[8] Rapp K. J., A. Sbravati, J. Vandermaar and M. Rave, "Evaluation of Transformer Components for High Temperature Transformers," *IEEE Electrical Insulation Conference* (EIC), San Antonio, TX, USA, 2018, pp. 385-389.

[9] Abdel Aleem S. H. E., A. F. Zobaa, M. E. Balci and S. M. Ismael, "Harmonic Overloading Minimization of Frequency-Dependent Components in Harmonics Polluted Distribution

Systems Using Harris Hawks Optimization Algorithm," *IEEE Access*, vol. 7, pp. 100824-100837, 2019.

[10] Nyandeni D. B., M. Phoshoko, R. Murray, B. A Thango, "Transformer Oil Degradation on PV Plants – A Case Study," *8th South African Regional Conference* (CIGRE), 14-17 November, 2017.

[11] Eeckhoudt S., S. Autru, L. Lerat, "Stray gassing of transformer insulating oils: impact of materials, oxygen content, additives, incubation time and temperature, and its relationship to oxidation stability," *IEEE Electrical Insulation Magazine*, Volume: 33, 2017.

[12] IEEE Std. C57.110™, *IEEE Recommended Practice for Establishing Liquid-Filled and Dry–Type Power and Distribution Transformer Capability When Supplying Nonsinusoidal Load Currents*, August, 2008.

[13] IEEE Std. C57.110™, *IEEE Recommended Practice for Establishing Liquid Immersed and Dry-Type Power and Distribution Transformer Capability when Supplying Nonsinusoidal Load Currents*, June, 2018.

[14] Thompson J. J., *On the Heat Produced by Eddy Currents in an Iron Plate Exposed to an Alternating Magnetic Field*, The electrician, April 1892.

[15] IEC 60076-7:2005 *Power transformers. Loading guide for oil-immersed power transformers*, December 2010.

[16] C57.91-*1995 - IEEE Guide for Loading Mineral-Oil-Immersed Transformers*, 1996.

Chapter 5

DEGREE OF POLYMERIZATION OF CELLULOSE INSULATION

In South Africa, transformers are the most critical links in the power system's generation, transmission, and distribution. Their insulation system is comprised of dielectric oil and cellulose paper. The ageing of the insulating system is influenced by several factors, including high loading profile, temperature rise, and short circuits. Many diagnostic tests can be performed, and correctional measures may be taken to guarantee that consumers receive an uninterrupted power supply. The remaining life of the transformer's cellulose paper insulation determines when the transformer's useful life ends. The mechanical or tensile strength of the solid insulation determines the ageing of the paper insulation, which is calculated in terms of its Degree of Polymerization (DP). The direct method, which involves actual paper samples and is thus expensive and invasive, is the conventional method for calculating DP. Recently, a non-intrusive indirect approach has been used to estimate the DP value from the furan compounds dissolved in the dielectric oil.

In this chapter, the impact of the DP on the remnant life of four different transformer cellulose insulating materials is studied. An analysis of various cellulose insulation DP models is employed in estimating the DP and consequently the remnant life. The chapter further proposes new formulae correlating 2-furaldehyde (2FAL) and DP in relation to the ageing process using regression analysis on a fleet of collected transformer oil samples. The results are compared with measured DP values and yield an estimation error of less than 3% and 4%, respectively.

1. INTRODUCTION

A transformer's proper operation is essential for the transmission and distribution of electricity in a power grid. The most effective use of these transformers ensures a cost-effective and efficient power supply [1-3]. As a result of deregulation, electric utilities are under pressure to reduce the cost of electricity production and are left with the option of using power apparatus that are close to their rated values. This rated and sometimes short overloading of transformers causes health problems and electrical and thermal stresses [4-6]. As a result, condition monitoring of transformers is important for maintaining and testing the health of transformers. According to the IEEE loading guide [7], the life of a transformer is typically estimated to be 25 to 30 years. However, due to robust condition monitoring and residual life assessment technique implementations [8-12], many transformers in the electricity board have been in operation for more than 30 years. These strategies help to avoid unplanned outages, which saves money in a power grid [13].

The majority of transformers have a dielectric oil and solid paper insulation system [14]. As several variables can affect the ageing process in a transformer, the state of this insulation defines the condition of each unit in service [15]. Overheating, long periods of service at high load, decay, moisture, high currents, and other factors all affect the ageing of the insulation system in transformers [16]. The most important causes of insulating device ageing are moisture and temperature of paper and oil, and oxidation accelerates the deterioration of insulation paper. Dielectric oil in transformers decomposes into hydrogen and low-molecular-weight gases like acetylene, ethane, ethylene, and methane, which can be detected using dissolved gas analysis [17–19]. Depending on various conditions that exist within a transformer or during its operation, the percentage of these gases varies from one transformer to the next. As these transformers' solid insulation (i.e., Kraft paper) decomposes, various items appear in the oil, including CO, CO_2, water, and small quantities of furan compounds [20]. This understanding of insulation decomposition enables the monitoring of different parameters to determine insulation and health degradation. Acidity, dielectric power, water content, gas content, loss factor, and furan concentration in transformer oil are all important indicators for determining a transformer's condition.

This chapter presents a summary review of the development of furan compounds and the interrelation between 2-FAL and DP. Additionally, a thorough examination of various DP models is presented which establishes the models describing the interrelation between 2-FAL and DP. The proposed new formulae based upon regression analysis on a fleet of transformer oil samples are also introduced. Further, a comparative study is carried out on the considered models by considering the 2-FAL data of four transformers to determine their DP and, wherefore the remnant life.

2. INTERRELATION BETWEEN 2-FAL AND DP

Solid insulation in transformers contains cellulose, which, when exposed to temperatures of 100°C or higher for different purposes, produces degradation by-products, some of which are soluble in oil [21-22]. These by-products are formed as a result of ageing and dissolve in oil, so this oil can be checked for the presence of furans. These are quickly sampled and can be used as markers of ageing. According to recent research [21-23], Furan compounds formed by electrical discharges affect cellulose, but only in a minor way. As these cellulosic materials of solid insulation are subjected to a very high temperature (i.e., 120°C & above), Furanic compounds in large quantities may be formed as a result of thermal stress or thermal ageing [21-22]. The rate at which these Furanic compounds shape can be influenced by a number of other factors, including water content and oxygen concentration. Furanic compound concentrations are measured in ug/L [21-23]. The majority of Furanic compounds are unstable in the eluent, despite the fact that there are many of them. There are five Furanic compounds that are widely used for various diagnostic procedures, and they are mentioned below as [23]:

- 2-Furaldehyde (2-FAL)
- 2-Acetylfuran (2ACF)
- 2-Furfuryl alcohol (2FOL)
- 5-Methylfurfural (5MF)
- 5-Hydroxymethylfurfural (5HF).

The stability of Furanic compounds is of greater importance because it helps in drawing conclusions from the study. The engineer's results would be distorted as a result of these unstable compounds. Some of the above-mentioned Furanic compounds become formed with age and are extremely unstable under various conditions. As a result, none of them is useful. According to laboratory tests performed by various researchers, 2-furaldehyde (2FAL), also known as 2 furfural of cellulose ageing in insulation systems, is the most stable by-product over time. As a result, it is commonly used to forecast the paper DP value.

Table 15. Primary cause for furan compounds

Furan compound	Primary cause
2-Furaldehyde	Overtemperature and regular ageing
2-Acetylfuran	Scarce, no conclusive trigger
2-Furfuryl alcohol	Elevated moisture content
5-Methylfurfural	Elevated temperatures
5-Hydroxymethylfurfural	Oxidation

3. INTERRELATION BETWEEN 2-FAL AND DP

According to recent research, indirect studies on oil can be carried out by examining the concentration of Furanic compounds produced during ageing. Furanic compounds migrate from paper to oil during the ageing process, and the DP value can be measured by analyzing this oil. The measurement of Furanic compounds in an oil sample is straightforward, but the interpretation is more difficult. As previously mentioned, several pathways are involved in the ageing process. Carbon-oxide gases and moisture are more dominant products at low temperatures; at moderate temperatures, these compounds are dominant; and at high temperatures, these compounds are extremely unstable. Several studies have looked into the ageing of paper and attempted to connect furans to the DP value. These studies focus on information gathered from various specimens collected through laboratory testing when transformers were out of operation for repairs.

In the current study, new formulae are developed based upon regression analysis on a fleet of transformer oil samples. Regression is a technique for estimating the value of one of two variables [x, y] using the values of the other variables. Simple linear regression, multiple linear regression, and nonlinear regression are the three forms of regression. As a consequence of the simple regression, the dependent variable y is dependent on one independent variable x, and their relationship is linear. A multiple linear regression occurs when the dependent variable y is dependent on more than one variable. Furthermore, the relationship between the dependent variable y and the other independent variables may be nonlinear in some situations.

The proposed logarithmic and polynomial experimental formulae in this chapter are then expressed in Eq. (1) and Eq. (2).

$$DP = -123{,}6 \times \ln 2FAL + 456{,}38 \tag{1}$$

The coefficient of determination (R^2) is a mathematical measurement that explores how a change in one variable affects another [12]. It's a measure of how powerful the interactions between two or more variables are. For Eq. (1), the computed coefficient of determination value is:

$$R^2 = 0.9809$$

The proposed polynomial is expressed as:

$$DP = 0{,}2358 \times 2FAL^6 - 5{,}4819 \times 2FAL^5 + 49{,}471 \times 2FAL^4 - 220{,}64 \times 2FAL^3 + 520{,}55 \times 2FAL^2 - 692{,}51 \times 2FAL + 811{,}15 \tag{2}$$

The corresponding coefficient of determination:

$R^2 = 0{,}9913$

Table 16. Models relating 2FAL and DP

Eq.	Author	DP experiential formula	Ref.
3	Chendong	$DP = \dfrac{\log_{10} 2FAL - 1{,}51}{-0{,}0035}$	[24]
4	Vaurchex	$DP = \dfrac{2{,}6 - \log_{10} 2FAL}{0{,}0049}$	[25]
1	Thango (Eq. 1)	$DP = -123{,}6 \times \ln 2FAL + 456{,}38$	
2	Thango (Eq. 2)	$DP = 0{,}2358 \times 2FAL^6 - 5{,}4819 \times 2FAL^5 + 49{,}471 \times 2FAL^4 - 220{,}64 \times 2FAL^3 + 520{,}55 \times 2FAL^2 - 692{,}51 \times 2FAL + 811{,}15$	

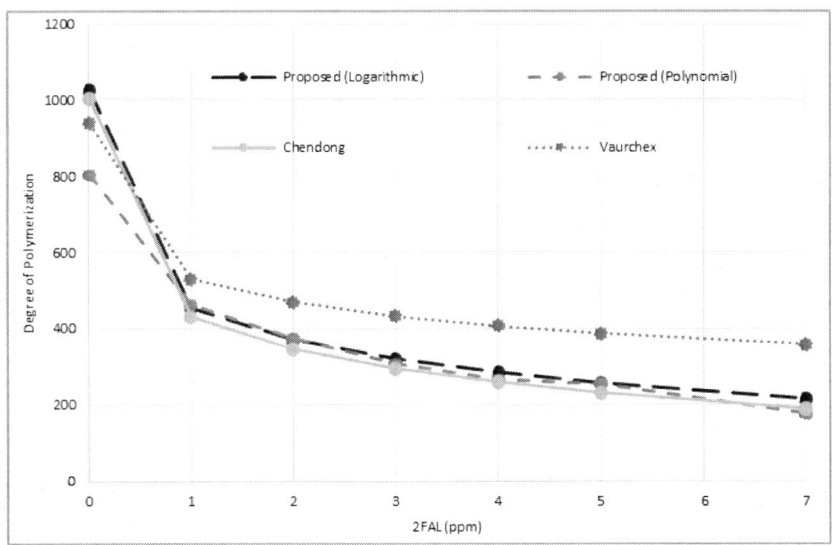

Figure 17. Performance of DP models at different 2FAL concentrations.

Aside from that, mathematical models based on the relationship between the furans (2-FAL) and the DP value have been established.

The following in Table 16 are the models that have been proposed by Chendong [24] and Vaurchex [25].

Vaurchex [25], this equation was developed based on the results of sealed ageing experiments of Kraft paper at elevated temperatures. Chendong [24], based on data collected from transformers with normal Kraft paper and free-breathing conservators, created this equation. 2-FAL is calculated in parts per million.

The performance of the models considered herein at different 2FAL concentrations is presented as shown in Figure 17 above.

4. REMNANT LIFE OF TRANSFORMER

Now, the percentage remaining life of the transformer is determined using the following equation, where the DP values range from 1200 ppm to 1000 ppm, with 200 ppm for degraded paper insulation.

The remaining service lifetime (RSL) of a unit in service and the percentage thereof, respectively, can be ascertained the Eq. (5) and Eq. (6) as shown in Table 17 below.

Table 17. Remaining service lifetime (RSL) formulae

Eq.	Author	RSL formula	Ref.
5	Pradhan	$RSL = 20{,}5 \times ln\left(\frac{1100}{2FAL}\right)$	[26]
6	Kanumuri	$\% RSL = \frac{LOG10(E63) - 2{,}903}{-0{,}006021}$	[27]

Eq. (6) gives the remaining life as a percentage, but to measure transformer paper life in years, Eq. (5) proposed by Pradhan et al. [26] may be employed. The parameters used by Pradhan et al. [17] assume initial DP (DP0) as 1100 and a 35-year transformer lifespan.

5. CASE STUDIES

In this section, the sample results of Furan Analysis of four transformers are presented as shown in Table 18. The DP has been measured for each of the case studies. At a glance, transformer (Trfr) 1 has the lowest 2FAL concentration with a slightly higher measured DP, which suggest this unit is older than the other units. Additionally, Trfr 4 has a higher 2FAL value and the lowest measured DP.

Table 18. Case scenarios

Trfr.	2FAL (ppm)	Measured DP
1	1,464	400
2	2,021	360
3	2,789	320
4	3,851	280

To substantiate the proposed models and to assess their performance against existing models, the DP of the presented case studies has been computed using the 2FAL concentrations (in ppm) and tabulated as shown in Table 19 below.

The error of estimate of the results in Table 19 are presented in Table 20 below to assess the performance of the models against the measured DP.

Table 19. Degree of polymerization

Trfr.	2FAL	Measured DP	Thango (Eq. 1)	Thango (Eq. 2)	Chendong	Vaurchex
1	1,464	400	409	413	384	497
2	2,021	360	369	373	344	468
3	2,789	320	330	321	304	440
4	3,851	280	290	270	264	411

Table 20. Degree of polymerization (Error estimate)

Trfr.	2FAL	Measured DP	Thango (Eq. 1)	Thango (Eq. 2)	Chendong	Vaurchex
1	1,464	400	2%	3%	-4%	24%
2	2,021	360	3%	4%	-4%	30%
3	2,789	320	3%	0%	-5%	37%
4	3,851	280	3%	-4%	-6%	47%

Thango (Eq. 1) and (Eq. 2) are observed to have an error of estimate of less than 3% and 4% in comparison to the measured DP. Chendong and Vaurchex have an error of estimation of 6% and 47%, respectively. In comparison to the measured DP and the Chendong model, the proposed models performed exceptionally well in evaluating these transformers.

Consequently, the remnant service life of the unit based upon the calculated DP is computed as shown in Table 21 below.

Table 21. Remnant life of transformer

Trfr.	2FAL	Measured DP	Thango (Eq. 1)	Thango (Eq. 2)	Chendong	Vaurchex
1	1,464	20,74	20,27	20,06	21,57	16,29
2	2,021	22,90	22,37	22,17	23,82	17,51
3	2,789	25,31	24,71	25,22	26,35	18,80
4	3,851	28,05	27,35	28,82	29,25	20,18

The error of estimate of the calculated remnant life of studied units is tabulated in Table 22 below.

The results yield an error of estimate of less than 2% and 3% respectively for Thango (Eq. 1) and (Eq. 2) while Chendong and Vaurchex indicate 4% and 26% respectively.

Using Eq. (6), these units' remaining percentage service lifetime was calculated as shown in Table 23 below.

Table 22. Remnant life of transformer (Error estimate)

Trfr.	2FAL	Thango (Eq. 1)	Thango (Eq. 2)	Chendong	Vaurchex
1	1,464	-2%	-3%	4%	-21%
2	2,021	-2%	-3%	4%	-24%
3	2,789	-2%	0%	4%	-26%
4	3,851	-2%	3%	4%	-28%

Table 23. Percentage Service Lifetime

Trfr.	2FAL	Measured DP	Thango (Eq. 1)	Thango (Eq. 2)	Chendong	Vaurchex
1	1,464	49,98	48,33	47,61	52,90	34,35
2	2,021	57,58	55,72	55,03	60,83	38,62
3	2,789	66,08	63,94	65,76	69,74	43,16
4	3,851	75,71	73,25	78,42	79,92	48,01

The mathematical models that may be used to measure the residual lifetime of transformers have been found to be not completely reliable to accurately estimate the DP condition. As a result, when the transformer is in service, it is pivotal to investigate the dielectric and physicochemical properties of the dielectric oil. Only recent sample measurements of oil properties cannot be used to determine the transformer state. Variables that are measured and deterioration products will not be in a constant pattern as the dielectric oil and insulating paper age and will fluctuate over time. The stability of 2-FAL present in oil in terms of temperature and time, the time it takes to find 2-FAL concentration after the oil is processed gives the ageing process and up to what extent the inaccuracy of indirect methods can be tolerated and why are some of the uncertainties that need to be investigated. The credibility of 2-FAL stability considers recent discoveries, and new models should be established by taking into account a greater number of variables that are highly accurate and effective in calculating the remaining life.

CONCLUSION

In this chapter, the results of mathematical models that show a correlation between the furan concentration(2-FAL) and the DP of the four different field transformers are analyzed, compared, and different conclusions are drawn based on the results obtained from the existing and proposed models. In addition, the formation or creation of various equations based on various laboratory findings and assumptions considered for each model is briefly discussed in this review. In this paper, a graph of the DP of cellulose insulation at different 2-FAL values has been demonstrated.

For power system operators, the stage at which a transformer must be replaced and its remaining lifetime are critical considerations. These questions are dependent on the state of the machines' insulating system. In reality, knowing the condition of the dielectric paper is critical for proper transformer fleet management. Since it is simple to take an oil sample, physicochemical analysis of oil has become the most widely used method for estimating transformer conditions. On the other hand, monitoring the condition of dielectric paper is more difficult. As a result, stopping the device and extracting the oil would be needed in order to obtain a sample of dielectric paper, which is not feasible. A direct association between furan content and paper condition has been discovered in various experimental laboratory studies. When the transformer age is high, finding a connection between furans and paper degradation becomes nearly impossible. Only in the early stages of the transformer's operation, using the formulas suggested in the literature, will it be possible to estimate the approximate condition of the paper from the furans content. It's also worth noting that oil replacement or regeneration makes it difficult to determine the true state of dielectric paper. Even if the insulating paper is nearing the end of its useful life, the oil can have optimal physicochemical and dielectric properties.

REFERENCES

[1] Liang S., "A high power and high efficiency PC power supply topology with low cost design to meet 80 Plus Bronze requirements," *2009 IEEE International Conference on Industrial Technology*, Churchill, VIC, Australia, 2009, pp. 1-6, doi: 10.1109/ICIT.2009.4939582.

[2] Hayashiya H. et al., "Cost impacts of high efficiency power supply technologies in railway power supply - Traction and Station -," *2012 15th International Power Electronics and Motion Control Conference* (EPE/PEMC), Novi Sad, Serbia, 2012, pp. LS3e.4-1-LS3e.4-6, doi: 10.1109/EPEPEMC.2012.6397441.

[3] Ghosh S., S. C. Basu and P. P. Sengupta, "Improvement of financial efficiency and cost effectiveness in energy sector: A case study from Indian thermal power plant," *2010 International Conference on Education and Management Technology*, Cairo, Egypt, 2010, pp. 512-516, doi: 10.1109/ICEMT.2010.5657605.

[4] Venkataswamy R., K. U. Rao and P. Meena, "Deformation Diagnostic Methods for Transformer Winding through System Identification," *2019 International Conference on Data Science and Communication (IconDSC)*, Bangalore, India, 2019, pp. 1-7, doi: 10.1109/IconDSC.2019.8816967.

[5] Sen P. K. and Sarunpong Pansuwan, "Overloading and loss-of-life assessment guidelines of oil-cooled transformers," *2001 Rural Electric Power Conference. Papers Presented at the 45th*

Annual Conference (Cat. No.01CH37214), Little Rock, AR, USA, 2001, pp. B4/1-B4/8, doi: 10.1109/REPCON.2001.949516.

[6] Clark L. W., "Substation-transformer emergency overloading practice," in *Electrical Engineering*, vol. 62, no. 3, pp. 126-132, March 1943, doi: 10.1109/EE.1943.6435652.

[7] "IEEE Guide for Loading Mineral-Oil-Immersed Transformers," in *IEEE Std C57.91-1995*, vol., no., pp.1-112, 30 Nov. 1995, doi: 10.1109/IEEESTD.1995.8684643.

[8] Montanari G. C., D. Fabiani, P. Morshuis and L. Dissado, "Why residual life estimation and maintenance strategies for electrical insulation systems have to rely upon condition monitoring," in *IEEE Transactions on Dielectrics and Electrical Insulation*, vol. 23, no. 3, pp. 1375-1385, June 2016, doi: 10.1109/TDEI.2015.005613.

[9] Zhao S., V. Makis, S. Chen and Y. Li, "Evaluation of Reliability Function and Mean Residual Life for Degrading Systems Subject to Condition Monitoring and Random Failure," in *IEEE Transactions on Reliability*, vol. 67, no. 1, pp. 13-25, March 2018, doi: 10.1109/TR.2017.2779322.

[10] Montanari G. C., P. Seri and R. E. Hebner, "A scheme for the Health Index and residual life of cables based on measurement and monitoring of diagnostic quantities," *2018 IEEE Power & Energy Society General Meeting (PESGM)*, Portland, OR, USA, 2018, pp. 1-5, doi: 10.1109/PESGM.2018.8585858.

[11] Einav I. and Q. Sun, "Revealing the stresses — New approach for industrial safety, reliability and residual life assessment," *2016 IEEE Far East NDT New Technology & Application Forum* (FENDT), NanChang, China, 2016, pp. 62-65, doi: 10.1109/FENDT.2016.7991996.

[12] Matsui T., Yasuo Nakahara, Kazuo Nishiyama, Noboru Urabe and Masayoshi Itoh, *Development of remaining life assessment*

for oil-immersed transformer using structured neural networks, 2009 ICCAS-SICE, Fukuoka, Japan, 2009, pp. 1855-1858.
[13] "IEEE Guide for Application for Monitoring Equipment to Liquid-Immersed Transformers and Components," in *IEEE Std C57.143-2012, vol.*, no., pp.1-83, 19 Dec. 2012, doi: 10.1109/IEEESTD.2012.6387561.
[14] Sbravati A., K. Rapp, P. Schmitt and C. Krause, "Transformer insulation structure for dielectric liquids with higher permittivity," *2017 IEEE 19th International Conference on Dielectric Liquids* (ICDL), Manchester, UK, 2017, pp. 1-4, doi: 10.1109/ICDL.2017.8124705.
[15] Koch M.and T. Prevost, "Analysis of dielectric response measurements for condition assessment of oil-paper transformer insulation," in *IEEE Transactions on Dielectrics and Electrical Insulation*, vol. 19, no. 6, pp. 1908-1915, December 2012, doi: 10.1109/TDEI.2012.6396947.
[16] Hribernik W., G. Pascoli and K. Frohlich, "An advanced model-based diagnosis system for online detection of the moisture content of power transformer insulations," *Conference Record of the 2008 IEEE International Symposium on Electrical Insulation*, Vancouver, BC, Canada, 2008, pp. 187-191, doi: 10.1109/ELINSL.2008.4570307.
[17] Hettiwatte S. N. and H. A. Fonseka, "Analysis and interpretation of dissolved gases in transformer oil: A case study," *2012 IEEE International Conference on Condition Monitoring and Diagnosis*, Bali, Indonesia, 2012, pp. 35-38, doi: 10.1109/CMD.2012.6416435
[18] Thang K. F., R. K. Aggarwal, A. J. McGrail and D. G. Esp, "Analysis of power transformer dissolved gas data using the self-organizing map," in *IEEE Transactions on Power Delivery*, vol. 18, no. 4, pp. 1241-1248, Oct. 2003, doi: 10.1109/TPWRD.2003.817733.

[19] Wang X. F., Z. D. Wang, Q. Liu, G. Wilson, P. Jarman and D. Walker, "Evaluation of mass transfer rate of dissolved gases in transformer oils," *2016 International Conference on Condition Monitoring and Diagnosis* (CMD), Xi'an, China, 2016, pp. 477-480, doi: 10.1109/CMD.2016.7757865.

[20] Thango B. A., J. A. Jordaan and A. F. nnachi, "Stray Gassing of Transformer Oil in Distributed Solar Photovoltaic (DSPV) Systems," *2020 6th IEEE International Energy Conference (ENERGYCon)*, Gammarth, Tunis, Tunisia, 2020, pp. 484-488, doi: 10.1109/ENERGYCon48941.2020.9236522.

[21] Abd El-Aal R. A., K. Helal, A. M. M. Hassan and S. S. Dessouky, "Prediction of Transformers Conditions and Lifetime Using Furan Compounds Analysis," in *IEEE Access*, vol. 7, pp. 102264-102273, 2019, doi: 10.1109/ACCESS.2019.2931422.

[22] Verma P., M. Roy, R. K. Tiwari and S. Chandra, "Generation of furanic compounds in transformer oil under accelerated thermal and electrical stress," *Proceedings Electrical Insulation Conference and Electrical Manufacturing Expo*, 2005., Indianapolis, IN, USA, 2005, pp. 112-116, doi: 10.1109/EEIC.2005.1566270.

[23] Cheim L., D. Platts, T. Prevost and S. Xu, "Furan analysis for liquid power transformers," in *IEEE Electrical Insulation Magazine*, vol. 28, no. 2, pp. 8-21, March-April 2012, doi: 10.1109/MEI.2012.6159177.

[24] Chendong X., F. Qiming, X. Shiheng, "To Estimate the Ageing Status of Transformers by Furfural Concentration in the Oil," In *Proceedings of the CIGRE Study Committee 33 Colloquium*, Leningrad, Moscow; 1991.

[25] Vaurchex H., I. Höhleim and A. J. Kachler, *"Transformer aging research on furanic compounds dissolved in insulating oil,"* CIGRE, 2002.

[26] Pradhan M. K. and T. S. Ramu, "On the estimation of elapsed life of oil-immersed power transformers," in *IEEE Transactions on Power Delivery*, vol. 20, no. 3, pp. 1962-1969, July 2005, doi: 10.1109/TPWRD.2005.848663.

[27] Kanumuri D., V. Sharma, O. P. Rahi, O P, "Analysis Using Various Approaches for Residual Life Estimation of Power Transformers," *International Journal on Electrical Engineering and Informatics*, Bandung Vol. 11, no. 2, (Jun 2019): 389-407. doi:10.15676/ijeei.2019.11.2.11.

Chapter 6

TRANSFORMER PREVENTIVE MAINTENANCE OF TRANSFORMER HEALTH INDEX THROUGH STRAY GASSING

Preventive maintenance endeavours in transformers aim to appraise the risk by means of the computation and surveillance of the health index of transformers in service. A prevalent method employed in the preventive maintenance and assessment of the health index of transformers is the Dissolved Gas Analysis (DGA). A contemporary trend is to utilize an online DGA monitoring instrument while carrying on to conduct investigations in the test laboratory. Most recently, the testing of stray gassing has only been conducted by oil suppliers. With the sharp increase in reports of peculiar stray gassing phenomena in relatively new transformers, typically not more than ten years in-service, transformer manufactures and utility owners began conducting this test in-house. Though an established testing procedure for stray gassing exists, there is still a lack of coherent criterion on the stray gassing limits and an appropriate period when stray gassing patterns should be a concern. In this paper, numerous DGA methods have been adopted to analyze the oil sample results of a 2000kVA transformer in-service that has been exhibiting stray gassing patterns. The objective of this paper is to distinguish the stray gassing phenomena of uninhibited mineral oil so that they are unambiguous from other occurring fault types. A wide-scale study on transformers filled with uninhibited mineral oils has been carried out to develop local stray gassing limits for transformers in the South African power grid, specifically in solar photovoltaic (PV) plants. Premier findings indicate that the Hydrogen gas constituted by stray gassing is often misinterpreted to corona partial discharges in the transformers.

1. INTRODUCTION

In order to meet the ever-increasing global energy demand, power technology has faced numerous challenges [1-2]. Despite this, oil-filled transformer technology has remained virtually unchanged for nearly a century. Currently, a large fleet of transformers are subjected to various accumulated stresses and are nearing the end of their projected lifespan [3-5]. As a result, many of these transformers would need to be replaced or repaired to ensure the power supply's safety. On the other hand, many power agencies are focusing their efforts on monitoring and diagnostic techniques that can provide a reliable assessment of the transformer's condition. Online monitoring techniques based on dissolved gas analysis (DGA) have gained much traction in this trend as a continuous and non-invasive process, but these monitoring techniques still have cost and technical limitations when it comes to diagnosing faults in transformers [6-8].

In many countries, the average age of transformer fleets has reached 30 years, which was the maximum lifespan built for these transformers for them to operate safely and efficiently [9-10]. Furthermore, to reduce service prices, deregulation of the electricity market has forced power companies to stop investing in new transformers and instead use their existing electrical components at maximum capacity for longer distances and times. As a result of the increased concern regarding the existing condition assessment of transformers, maintenance measures and condition monitoring techniques have been introduced. Condition techniques may detect defective conditions in transformers, extending their technical lifetime and preventing costly breakdowns [11].

Under the current stringent operational requirements of solar PVs, transformers are highly susceptible to developing faults that can result in high outage costs, as well as fire or explosions [12]. High temperatures, strong electrical fields, electrical discharges, mechanical stresses, insulation damage, and pollutants can all expose a transformer to imminent and irreversible damage, regardless of its age [13]. Due to their excellent dielectric strength and low cost, the combination of transformer oil and cellulose paper has remained the most efficient insulating device to date [14]. On the other hand, the transformer oil is oxidized as a result of free radical reactions between unstable hydrocarbon molecules and oxygen [14]. The presence of copper and iron in the transformer catalyzes these reactions, which are then intensified by the heat dissipated from the windings and core. Oil oxidation results in a loss of dielectric strength and the formation of a variety of solid, liquid, and gaseous compounds. These compounds can also contribute to the creation of transformer faults.

The oil in a transformer is subjected to a slow continuous oxidation phase under normal operating conditions. However, due to a higher-than-usual amount of energy dissipated by an electrical or thermal fault in the transformer, the oxidation process may accelerate suddenly, resulting in a sudden increase in the concentrations of some hydrocarbon gases and other compounds [13-14]. Thus, the condition of the transformer oil, which includes dissolved gas concentrations, sludge particle appearance, colour, and other factors, offers valuable information about the transformer's technical state and malfunctioning.

In this chapter, the oil sample results of a 2000kVA transformer in-service exhibiting stray gassing patterns are analyzed. This chapter aims to distinguish the stray gassing phenomena of uninhibited mineral oil so that they are unambiguous from other occurring fault types by developing local stray gassing limits for transformers in the South African power grid, specifically in solar PV plants.

2. Dissolved Gas Analysis

To track and detect electrical or thermal faults, the guides [15-17] recommend dissolved gas concentration limits in transformer oil. Table 24 [15] demonstrate these limitations typical gas concentration (95th percentile) as a function of O_2/N_2 and age in µL/L (ppm). The reference limits were defined for transformers of specific types in specific installation and location conditions. As a result, an electric utility's transformer maintenance can be justified within the bounds of the requirements. It cannot, however, be completely realized by strictly adhering to these guidelines. The standard rates of gas increase presented in the guides [15-16] are shown in Table 25, display the acetylene values as a function of the on-load tap-changer (OLTC). The typical concentration and annual increase of acetylene in power transformers without OLTC or without communicating OLTC are lower than in power transformers with communicating OLTC [15-16].

The guidelines [17-18] allow electric utilities to set their own thresholds for gas concentrations and annual rises in gas concentrations. A method for calculating these limits is proposed in subsection A, specifically for solar PVs. A fleet of solar PV transformers and their operational conditions are considered in this criterion.

Table 24. Typical gas concentration (95th percentile) as a function of O_2/N_2 and age in μL/L (ppm)

Gas	O_2/N_2 Ratio ≤ 0.2				O_2/N_2 Ratio >0.2			
	Transformer Age (in Years)				Transformer Age (in Years)			
	Unknown	1-9	10-30	>30	Unknown	1-9	10-30	>30
Hydrogen (H_2)	200	200			90	90		
Methane (CH_4)	150	100	150	200	50	60		30
Ethane (C_2H_6)	175	70	175	250	40	30	40	
Ethylene (C_2H_4)	100	40	175	90	100	80	125	
Acetylene (C_2H_2)	2	2	4		7	7		
Carbon monoxide (CO)	1100	1100			600	600		
Carbon dioxide (CO_2)	12500	7000	14000		7000	3500	5000	8000

Table 25. Typical gas concentration (90th percentile)

Gas	Not OLTC	Communicating OLTC
Hydrogen (H_2)	60-150	75-150
Methane (CH_4)	40-110	35-130
Ethane (C_2H_6)	50-90	0
Ethylene (C_2H_4)	60-280	110-250
Acetylene (C_2H_2)	3-50	80-270
Carbon monoxide (CO)	540-900	400-850
Carbon dioxide (CO_2)	5100-13000	5300-12000

Table 25 shows the acetylene values as a feature of the OLTC. The typical concentration and annual increase of acetylene in power transformers without OLTC or without communicating OLTC are lower than in power transformers with communicating OLTC.

A test procedure in which a transformer is tested at relatively short intervals (e.g., months, weeks, or days) to detect and characterize any gas formation that may occur and provide early notice of a rapidly deteriorating condition. This protocol is used to assess fault severity and confirm fault operation. It may also be used when the state of the device changes, such as after a unit relocation or a load rise. In [15], the following dissolved gas concentration surveillance of oil samples are tabulated as shown in Table 26.

Table 26. Gas concentration surveillance

Gas	Normal	Caution	Warning
Hydrogen (H_2)	<100	100-700	>700
Methane (CH_4)	<120	120-400	>400
Ethane (C_2H_6)	<65	65-100	>100
Ethylene (C_2H_4)	<50	50-100	>100
Acetylene (C_2H_2)	<2	2-5	>5
Carbon monoxide (CO)	<350	350-570	>570
Carbon dioxide (CO_2)	<700	700-1900	>1900

The DGA results may be used to identify the apparent fault form if irregular gas generation is verified. Gas increments or rates of gas formation, in addition to fault detection, suggest the fault's relative severity. Changes in fault type increments, rates, or evolution (e.g., from a lower energy fault type to a higher energy fault type) may indicate fault process worsening or moderation. Unless the gassing activity is so severe that immediate shutdown is needed, it may be prudent to take DGA samples more regularly and track for long enough to evaluate gas formation rates, determine whether the abnormality is transient, sporadic, or ongoing, and determine whether the fault type and intensity appear to be evolving. If it is important to keep a gassing transformer running for operational purposes, more frequent DGA surveillance may provide early warning of increased gas generation, which could mean that the transformer's condition is deteriorating.

The California State University Sacramento (CSUS) proposed the following guidelines in Table 27 to evaluate dissolved gas levels in power transformers [17]. The types of faults associated with gas concentration levels are as indicated in the interpretation below.

Table 27. CSUS Gas concentration surveillance

Gas	Normal	Elevated	Abnormal	Interpretation
H_2	<150	150-1000	>1000	Arcing, Corona
CH_4	<25	25-80	>80	Sparking
C_2H_6	<10	10-34	>35	Local overheating
C_2H_4	<20	20-100	>100	Severe overheating
C_2H_2	<15	15-70	>70	Arcing
CO	<500	50-1000	>1000	Severe overloading
CO_2	<1000	10000-15000	>15000	Severe overloading
TCDG	<720	720-5000	>15000	TCDG limit

3. PROPOSED STRAY GASSING LIMITS

In spite of the fact that DGA is widely known, there is an information gap of released data in addition to that on the guide, specifically for transformers intended to be of service to solar PVs.

This chapter used mineral oil sample results of a fleet of transformers in the South African power grid for distinguishing the stray gassing patterns of uninhibited mineral oil so that they are unambiguous from other occurring fault types to develop local stray gassing limits.

Table 28. Proposed stray gassing limits in μL/L (ppm)

Gas	Normal	Elevated	Interpretation
H_2	250-1000	>1000	Corona, PD
CH_4	50-100	>100	Low-energy density arcing
C_2H_6	10-40	>40	Low temperature thermal
C_2H_4	10-60	>60	High temperature thermal
C_2H_2	10-70	>70	Arcing, Corona
CO	10-900	>900	Thermal Overload >600°C
CO_2	10-13000	>13000	Thermal Overload >600°C

The methodology employed in this chapter is composed of three steps.

3.1. Step 1: Data Acquisition

The initial step involves collecting data samples of mineral oil-based transformers exclusively serving the solar PV environment. Next, the dissolved gases data is collated for analysis.

3.2. Step 2: Classification

The stray gassing limits are developed using natural breaks classification based on organic populations inherent in the collated sample results.

3.3. Step 3: Determining Sampling Frequency

Based on the actual condition of the units and their observed stray gassing patterns, the sampling frequency of the gasses is recommended.

A premier finding of this chapter indicates that the Hydrogen gas constituted by stray gassing is often misinterpreted to corona and partial discharges in the transformers. Additionally, the stray gassing patterns are linked up to the hotspot temperature due to distorted harmonics loads. Based on the peculiar operating conditions of solar PV applications and the condition of the units after testing, the study concluded that hydrogen concentrations between 250-1000 as normal ageing for this application. Subsequent sample results indicating an increase in a specific gas concentration imply an increase in the fault that has been interpreted by 27. The chapter proposes a standard sampling frequency of solar PV units for the utility owners to exercise caution is 4 to 6 months.

4. CASE STUDY

A 2000kVA mineral oil-immersed transformer that is facilitating a solar PV plant is considered for DGA studies. The unit was reported after it exhibited anomalies in the concentration of the dissolved gases. Generally, for transformers, an annual sample frequency would be a recommended practice. However, the operation conditions of solar PVs are unique and stray gassing limits of conventional units may not be applicable. Table 29 shows a regular sample taken on the 20th of September 2019 and subsequent samples taken on the 10th of November 2019 and 30th March 2020 to monitor the observed gassing activity.

The replacement of the transformer oil had a major effect on the gassing patterns, as shown in Table 29, suggesting oil degradation in the first samples collected. After the oil was replaced, the units was constantly tracked, and it was discovered that gas levels continued to rise in the unit.

In the next section, a comparative study of DGA methods is carried out to evaluate the status of the reported dissolved gases above.

Table 29. DGA combustible gas concentrations in μL/L (ppm)

Sample dates	20-Sept-2019	10-Nov-2019	30- Mar- 2020
H_2	1561	1131	950
CH_4	416	188	230
C_2H_6	618	388	320
C_2H_4	12	0	0
C_2H_2	0	0	0
CO	32	25	59
CO_2	964	856	1459
O_2	26188	24792	51005
N_2	68660	68402	151394
TCDG	2639	1732	1559

4.1. Observed Gas Concentration as a Function of O_2/N_2 and Age in μl/L (PPM)

The studied unit has been in-service not more than 10 years, so Table 24 $O_2/N_2 \leq 0.2$ and the 1-9 age in years category is used for the evaluation. The sample results from the 30th March 2020 are evaluated and tabulated as follows in Table 30.

Table 30. Observed 95th percentile gas concentrations in μL/L (ppm)

Sample dates	Transformer age Limit	30- Mar- 2020
H_2	200	950
CH_4	100	230
C_2H_6	70	320
C_2H_4	40	0
C_2H_2	2	0
CO	1100	59
CO_2	12500	1459

The transformer oil was replaced in the studied unit due to the pronounced concentrations of Hydrogen (H_2), Methane (CH_4) and Ethane (C_2H_6), which suggest corona, partial discharge and local overheating. A new DGA trend was developed for the units to see an appreciable, detectable difference in performance.

4.2. Observed 90% Concentration Values

The samples result analysis using the typical gas concentration recommended by IEEE Std. C57.104-2019 [15] in showed in Table 25 are tabulated below in Table 31.

The results indicate elevated levels of Hydrogen (H_2), Methane (CH_4) and Ethane (C_2H_6), which suggest corona, partial discharge and local overheating.

Table 31. Observed concentration values

Sample dates	30- Mar- 2020	Communicating OLTC	Above Limit
H_2	950	75-150	Yes
CH_4	230	35-130	Yes
C_2H_6	320	0	Yes
C_2H_4	0	110-250	No
C_2H_2	0	80-270	No
CO	59	400-850	No
CO_2	1459	5300-12000	No

4.3. Analysis Based on the Gas Surveillance Guidelines

Analysis based upon the Gas Surveillance guideline is presented in recommended by IEEE Std. C57.104-2019 [15] as shown in TABLE 26 has been interpreted as shown in Table 32 below.

Table 32. Gas concentration surveillance

Gas	Normal	Caution	Warning	30- Mar- 2020
H_2	<100	100-700	>700	950
CH_4	<120	120-400	>400	230
C_2H_6	<65	65-100	>100	320
C_2H_4	<50	50-100	>100	0
C_2H_2	<2	2-5	>5	0
CO	<350	350-570	>570	59
CO_2	<700	700-1900	>1900	1459

The TCG column's colour legend corresponds to the transformer operating conditions mentioned in 26. The sampled unit are pronounced for Hydrogen (H_2), Methane (CH_4) and Ethane (C_2H_6), which suggest corona, partial discharge and local overheating.

4.4. CSUS Gas Concentration Surveillance

Consequently, the sample results are analyzed using the CSUS gas concentration surveillance in Table 33 below.

Table 33. CSUS Gas concentration surveillance

Gas	Normal	Elevated	Abnormal	30- Mar-2020	Interpretation
H_2	< 150	150-1000	>1000	950	Arcing, Corona
CH_4	<25	25-80	> 80	230	Sparking
C_2H_6	<10	10-34	>35	320	Local overheating
C_2H_4	<20	20-100	>100	0	Severe overheating
C_2H_2	<15	15-70	>70	0	Arcing
CO	<500	50-1000	>1000	59	Severe overloading
CO_2	<700	720-5000	>15000	1459	Severe overloading
TCDG	<10000	10000-15000	>5000	1559	TCDG limit

TDCG levels at 1559 are classified as Condition 2 according to IEEE Std. C57.104-2019 [15]. The high level of Hydrogen and Methane indicates possible corona inside the unit.

4.5. Proposed Stray Gassing Limits

The sample results are analyzed using the proposed stray gassing limits as presented in Table 34 below.

High levels of Ethane are indicative of possible local hotspot temperature rise of the unit as a result of low-energy density arcing.

Hydrogen and Ethane gasses showed a decrease in production rates. Methane indicated an increase in production but is still below the normal ml/day rate. Hydrogen gas constituted by stray gassing is often misinterpreted to corona and partial discharges in the transformers. Additionally, the stray gassing patterns are linked up to the hotspot temperature due to distorted harmonics loads. Based on the peculiar operating conditions of solar PV applications and the condition of the units after testing, the study concluded that hydrogen concentrations between 250-1000 as normal ageing for this application.

Table 34. Proposed stray gassing limits in μL/L (ppm)

Gas	Normal	Elevated	30- Mar- 2020	Interpretation
H_2	250-1000	>1000	950	Corona, PD
CH_4	50-100	>100	230	Low-energy density arcing
C_2H_6	10-40	>40	320	Low temperature thermal
C_2H_4	10-60	>60	0	High temperature thermal
C_2H_2	10-70	>70	0	Arcing, Corona
CO	10-900	>900	59	Thermal Overload >600°C
CO_2	10-13000	>13000	1459	Thermal Overload >600°C

CONCLUSION

In this chapter, the sample results of a fleet of transformers exclusively servicing solar PV application has been studied for stray gassing analysis. The chapter contemplates transformers filled with uninhibited mineral-based oil. According to the findings discussed in this paper, the key factors affecting stray gassing patterns are the hotspot temperature of the active part components.

According to the study in contrast to other combustible gases, oil in contact with oxygen appears to contain higher amounts of hydrogen. The premier findings of this chapter indicate that the Hydrogen gas constituted by stray gassing is often misinterpreted to corona and partial discharges in the transformers. Additionally, the stray gassing patterns are linked up to the hotspot temperature due to distorted harmonics loads. Based on the peculiar operating conditions of solar PV applications and the condition of the units after testing, the study concluded that hydrogen concentrations between 250-1000 as normal ageing for this application.

REFERENCES

[1] Andrade J. V. B., R. S. Salles, M. N. S. Silva and B. D. Bonatto, "Falling Consumption and Demand for Electricity in South Africa - A Blessing and a Curse," *2020 IEEE PES/IAS PowerAfrica*, Nairobi, Kenya, 2020, pp. 1-5, doi: 10.1109/PowerAfrica49420.2020.9219878.

[2] Olanrewaju O. A., "Analysing Impacts Responsible for South Africa's Energy Consumption: LMDI Application," *2019 IEEE International Conference on Industrial Engineering and Engineering Management* (IEEM), Macao, China, 2019, pp. 1593-1596, doi: 10.1109/IEEM44572.2019.8978872.

[3] Nichols L. C., "Effect of overloads on transformer life," in *Electrical Engineering*, vol. 53, no. 12, pp. 1616-1621, Dec. 1934, doi: 10.1109/EE.1934.6540128.

[4] Thango B. A., J. A. Jordaan and A. F. Nnachi, "Effects of Current Harmonics on Maximum Loading Capability for Solar Power Plant Transformers," *2020 International SAUPEC/*

RobMech/PRASA Conference, Cape Town, South Africa, 2020, pp. 1-5, doi: 10.1109/SAUPEC/RobMech/ PRASA48453.2020. 9041101.

[5] Khederzadeh M., "Transformer Overload Management and Condition Monitoring," *Conference Record of the 2008 IEEE International Symposium on Electrical Insulation*, Vancouver, BC, Canada, 2008, pp. 116-119, doi: 10.1109/ELINSL.2008. 4570292.

[6] Zhou D. et al., "Examining acceptable Dissolved Gas Analysis level of in-service transformers," *2012 International Conference on High Voltage Engineering and Application*, Shanghai, China, 2012, pp. 612-616, doi: 10.1109/ICHVE.2012.6357050.

[7] Thango B. A., J. A. Jordaan and A. F. nnachi, "Stray Gassing of Transformer Oil in Distributed Solar Photovoltaic (DSPV) Systems," *2020 6th IEEE International Energy Conference (ENERGYCon)*, Gammarth, Tunis, Tunisia, 2020, pp. 484-488, doi: 10.1109/ENERGYCon48941.2020.9236522.

[8] Liang Y., Y. Xu, X. Wan, Y. Li, N. Liu and G. Zhang, "Dissolved gas analysis of transformer oil based on Deep Belief Networks," *2018 12th International Conference on the Properties and Applications of Dielectric Materials* (ICPADM), Xi'an, China, 2018, pp. 825-828, doi: 10.1109/ICPADM. 2018.8401156.

[9] Diwyacitta K., R. A. Prasojo, Suwarno and H. Gumilang, "Effects of lifetime and loading factor on dissolved gases in power transformers," *2017 International Conference on Electrical Engineering and Computer Science* (ICECOS), Palembang, Indonesia, 2017, pp. 243-247, doi: 10.1109/ ICECOS.2017.8167142.

[10] Okabe S., G. Ueta and T. Tsuboi, "Investigation of aging degradation status of insulating elements in oil-immersed transformer and its diagnostic method based on field measurement data," in *IEEE Transactions on Dielectrics and Electrical Insulation*, vol. 20, no. 1, pp. 346-355, February 2013, doi: 10.1109/TDEI.2013.6451376.

[11] Nyandeni D. B., M. Phoshoko, R. Murray, B. A. Thango, "Transformer Oil Degradation on PV Plants – A Case Study," *8th South Africa regional conference*, 14-17 November 2017.

[12] Thango B. A., J. A. Jordaan and A. F. Nnachi, "Contemplation of Harmonic Currents Loading on Large-Scale Photovoltaic Transformers," *2020 6th IEEE International Energy Conference* (ENERGYCon), Gammarth, Tunis, Tunisia, 2020, pp. 479-483, doi: 10.1109/ENERGYCon48941.2020.9236514.

[13] Zhang C. h. and J. M. K. Macalpine, "Furfural Concentration in Transformer Oil as an Indicator of Paper Ageing, Part 1: A Review," *2006 IEEE PES Power Systems Conference and Exposition*, Atlanta, GA, USA, 2006, pp. 1088-1091, doi: 10.1109/PSCE.2006.296461.

[14] Li S., Z. Ge, A. Abu-Siada, L. Yang, S. Li and K. Wakimoto, "A New Technique to Estimate the Degree of Polymerization of Insulation Paper Using Multiple Aging Parameters of Transformer Oil," in *IEEE Access*, vol. 7, pp. 157471-157479, 2019, doi: 10.1109/ACCESS.2019.2949580.

[15] "IEEE Guide for the Interpretation of Gases Generated in Mineral Oil-Immersed Transformers," in *IEEE Std C57.104-2019 (Revision of IEEE Std C57.104-2008), vol., no.*, pp.1-98, 1 Nov. 2019, doi: 10.1109/IEEESTD.2019.8890040.

[16] "IEEE Guide for the Interpretation of Gases Generated in Oil-Immersed Transformers," in *IEEE Std C57.104-1991*, vol., no., pp.0_1-, 1992, doi: 10.1109/IEEESTD.1992.106973.

[17] *IEC*. Mineral Oil-Filled Electrical Equipment in Service—Guidance on the Interpretation of Dissolved and Free Gases Analysis; IEC 60599:2015; IEC: Geneva, Switzerland, 2015.

[18] IEEE. *IEEE Guide for the Interpretation of Gases Generated in Mineral Oil-Immersed Transformers*; IEEE Std. C57.104-2019 (Revision of IEEE Std C57.104-2008); IEEE: Piscataway, NJ, USA, 2019; pp. 1–98.

Chapter 7

A DIAGNOSTIC STUDY OF DISSOLVED GASES IN TRANSFORMERS BASED ON FUZZY LOGIC APPROACH

Stray gassing of transformer oil is a convoluted physical phenomenon in which several parameters act simultaneously, thus making the understanding of Dissolved Gas Analysis (DGA) more daunting. Historically, prudent maintenance strategies are better conceived by studying the condition of units in service. Transformer condition assessment is critical to alleviating transformer management and decision matrix using unfailing, noninvasive diagnostics and monitoring mechanisms. The diagnosis of stray gassing of oil can be performed by distinguishing and embedding transformer severities using Fuzzy logic (FL) mapping.

Despite the fact that the IEC FL approach is handy in the appraisal of dissolved gases, the DGA results are impossible to match by current codes and, consequently, provide inconclusive results. Additionally, there is no quantitative measure for the possibility of distinct faults in the IEC FL approach. This chapter reveals a supplemental gas ratio C_2H_6/CH_4 based upon trial and error on a fleet of transformers. These supplemental ratios provide more detailed information about the thermal decomposition of transformer oil from various infancy faults and associated degradation gases transforming from Methane (CH4) to Ethane (C2H6) to Ethylene (C2H4). Further, this chapter provides an incipient fault diagnosis mechanism and some guidelines for effective planning and maintenance practice.

1. INTRODUCTION

The dielectric breakdown of transformer oil during service releases gas bubbles. This occurrence may potentially cause an undue default rate in oil-immersed transformers and is a matter of grave concern for both transformer manufacturers and utility owners. The dispersion of dissolved gases helps to unmask the nature of fault; and the associated gas ratios specify the fault's ferocity in the transformer. A swift and classical diagnostic procedure for incipient fault spotting is transformer oil's Dissolved Gas Analysis (DGA). In practice, oil samples are taken during factory acceptance tests (FATs) and during site routine testing for laboratory analysis and the parts per million (PPM) rates for individual gases within the oil samples are evaluated. Under normal operating conditions, the transformer oil releases dissolved gases containing Hydrogen (H_2), Methane (CH_4), Ethylene (C_2H_4), Acetylene (C_2H_2), and Ethane (C_2H_6). In the event that a fault is encountered, certain gases are generated rather in increased amounts than in normal transformer operating conditions. In diagnosing the nature of transformer fault, there are interpretive techniques available viz. three gas ratio (3GR) [1], Roger's ratio (RR) [2], Duval Triangle [3] and IEC ratio [4]. In the 3GR, the principal hydrocarbon dissolved gases concentrations are kept under surveillance. The gas concentrations are presented in parts per million (ppm). Recent work on the application of 3GR is carried out by the authors in [5, 6, 7 and [8]. In the RR approach (RRA), transformer faults are diagnosed by a coding scheme based upon dissolved gas ratios' ranges. In essence, the RR method is founded on thermal ageing principles. The gas analysis of oil-immersed transformers has been carried by various authors in [9, 10, 11] and [12]. A major drawback of the RR method is that it does not consider the gasses below permissible gas concentration levels, which generally results in erroneous interpretation of the results. The Duval Triangle (DT) method employs three dissolved gases and respective positioning in the Duval triangle by converting them into triangular coordinates. In [13], Permana et al. conducted a DGA study based upon the DT method on oil-immersed transformers facilitating geothermal power plants. The results indicated partial discharge (PD) and thermal faults due to surges in C_2H_2 and carbon dioxide (CO_2) activities on the oil data samples collected over the course of 3 years. In [14], in his DT study, Singh et al. tested over 100 defective transformers. The DT results indicated an accuracy of about 95% when compared to other diagnostic methods. At large, this journal reports that a single concentration level of the three gases can give an immediate fault diagnosis. Studies in [15] and [16] attempt to modify and automate the computerization of the DT method employing tools such as MATLAB.

In the case of multiple faults occurring simultaneously within the transformer, the dissolved gases are mixed up, ensuing perplexing gas ratios amongst various gas components. In cases of this nature, the Fuzzy logic approach (FLA) may be the most useful [16, 17, 18, 19] and [20]. This chapter intends to propose an FLA in which overcomes some of the shortcomings mentioned above. The FLA is established upon a large number of oil samples as primary data to propose an incipient fault diagnosis mechanism further. In order to test the performance of the proposed method, a comparative study is carried out against the RRA and the IEC FLA.

2. DIAGNOSTIC TECHNIQUES USING GAS RATIO APPROACH

Transformer faults are mainly recognizable whenever the cellulose and liquid insulation materials' permissible electrical and thermal values are exceeded. Considering that transformer infancy faults can be classified into electrical and thermal faults, the classification of these faults revolves around particular dissolved gases. In this section, diagnostic tools based upon three gas ratios are explored. Initially, a comprehensive application of the RRA is presented. Additionally, the IEC based FLA is also presented. At large, this section then presents a fault diagnosis technique based on the FLA using DGA samples collected over a fleet of transformers found to be exhibiting various peculiar gassing characteristics.

2.1. Roger Ratio Approach

The RRA employs three gas ratios comprising of five dissolved gases viz. CH_4/H_2, C_2H_4/C_2H_6 and C_2H_2/C_2H_4 [2, 3] and [4]. The spectrum of all gas ratios indicating a distinctive fault is tabulated as shown in Table 35.

Table 35. Failure mechanism of test transformers

Gas Ratio	Codes			
	0	1	2	5
$X = CH_4/H_2$	$0.1 \leq X \leq 1$	$1 \leq X \leq 3$	$X > 3$	$W < 0.1$
$Y = C_2H_4/C_2H_6$	$Y < 1$	$1 \leq Y \leq 3$	$Y > 3$	-
$Z = C_2H_2/C_2H_4$	$Z < 0.1$	$0.1 \leq Z \leq 3$	$Z > 3$	-

In normal operating conditions of transformers, the various dissolved gases considered above evolve. The various transformer fault mechanisms materialize when these gases exceed normally acceptable levels. Table 36 gives the various fault mechanisms that can be identified by the use of RRA.

Table 36. Incipient fault diagnosis mechanism

Fault code			Failure Mechanism
X	Y	Z	
0	0	0	No Fault
5	0	0	PD - Low energy density arcing
0	2	1	PD - High energy discharge
0	1	0	Low thermal fault
1	1	0	Thermal Fault <700°C
1	2	0	Thermal Fault>700°C

A major shortcoming of the RRA is that dissolved gas ratios may indicate a fault that may not conform to any specific fault type, necessitating other approaches to be employed for such instances.

2.2. Fuzzy Logic Approach: IEC

In the transformer industry, the IEC FLA codes are prevalent in interpreting the DGA oil samples. This approach has been standardized in various revisions of the IEC standards [2, 3] and [4]. The dissolved gas ratios, associated codes and limitations are presented in Table 37. These codes are then assigned to related fault mechanism.

Table 37. Failure mechanism of test transformers

Gas Ratio	Codes		
	0	1	2
$X = C_2H_2/C_2H_4$	$X < 0.1$	$0.1 < X < 3$	$X > 3$
$Y = CH_4/H_2$	$0.1 < Y < 1$	$Y < 0.1$	$Y > 1$
$Z = C_2H_4/C_2H_6$	$Z < 1$	$1 < Z < 3$	$Z > 3$

In Table 38, the corresponding fault codes from the gas ratios X, Y and Z above are presented.

Table 38. Incipient fault diagnosis mechanism

Fault code			Failure Mechanism
X	Y	Z	
0	0	0	No Fault
0	1	0	PD of low energy density
1	1	0	PD of high energy density
1 or 2	0	1 or 2	Discharges of low energy
1	0	2	Discharges of high energy
0	0	1	Thermal fault of low temp, <150°C
0	2	0	Thermal fault of low temp, 150-300°C
0	2	1	Thermal fault of medium temp, 300-700°C
0	2	2	Thermal fault of high temp, >700°C

The dissolved gas codes serves to easily formulate computer aided programs for fault detection.

2.3. Fuzzy Logic Approach: Proposed

Even though the FLA by IEC provides some beneficial DGA information, from a comparative study of diagnostic findings on over 300 units conducted through a local transformer manufacturer, it became apparent that the diagnostic precision was restricted to 70-80%. This suggests that there is, therefore a need to enhance the IEC FLA for gas analysis further. From classical methods such as Roger's and Dornenburg's, supplemental gas ratios have been employed. These supplemental ratios do provide more detailed information about the thermal decomposition of transformer oil from various infancy faults and associated degradation gases transforming from Methane (CH_4) to Ethane (C_2H_6) to Ethylene (C_2H_4).

In the efforts to further refine the symptomatic accuracy of the IEC FLA, this chapter reveals a supplemental gas ratio C_2H_6/CH_4 in the proposed FLA based upon hands-on experience. Below Table 39 is proposed to produce a supplemental gas ratio (GR) into the classical IEC 3 GR FLA.

Table 39. Failure mechanism of test transformers

Gas Ratio	Codes		
	0	1	2
$X = C_2H_6/CH_4$	$X < 0.1$	$0.1 < X < 2$	$X > 2$
$Y = CH_4/H_2$	$0.5 < Y < 1$	$Y < 0.5$	$Y > 1$
$Z = C_2H_4/C_2H_6$	$Z < 1$	$1 < Z < 3$	$Z > 3$

Based upon the fleet of transformers tested using the proposed FLA and other diagnostic methods, the incipient fault diagnosis mechanism in Table 40 below is proposed.

The proposed FLA substitute the classical logic rage for CH_4/H_2 to allow for more accuracy in the identification of infancy degradation gases. The proposed codes above close the gap for the shortcomings of RRA and IEC FLA because in many instances, the dissolved gas ratios indicate a fault that does not conform to any specific code and failure mechanism, resulting in many unknown diagnoses of many DGA oil samples.

Table 40. Incipient fault diagnosis mechanism

Gas Ratio			Failure Mechanism
X	Y	Z	
0	0	0	Normal aging
0	0	1	HST rise in metallic parts, <130°C
0	1	0	PD of low energy density
1	0	0	Increase in oil temp. 200-300°C
0	0	2	Continuous arcing with low energy density
0	2	0	Overheated core due to Radial magnetic flux leakage
2	0	0	Increase in oil temp. >300°C
0	1	1	HST rise in the winding conductors due to Eddy currents
1	0	1	Continuous arcing 200-300°C
1	1	0	Surface tracking of cellulose insulation
0	1	2	Damage to A or C Phase HV winding
0	2	1	HST rise in the winding conductors due to Eddy currents
1	0	2	Discharges of low energy between windings
1	2	0	Increase in temp. 150-200°C
2	0	1	Discharges of high energy
2	1	0	Damage to Interturn
0	2	2	Excessive heating of leads to tank <900°C
2	0	2	Continuous arcing >300°C
2	2	0	Wet cellulose insulation
1	1	1	Interturn insulation surface discharge activity
1	1	2	Oxidation <130°C
1	2	1	Increase in temp. >200°C
2	1	1	Continuous arcing >400°C
1	2	2	Excessive heating of leads to tank >900°C
2	1	2	Oil breakdown between cellulose insulation
2	2	1	Excessive vibrations
2	2	2	Mechanical failure

3. CASE STUDIES

To investigate the operational state of transformers during their service lifetime, the gas analysis of 10 oil samples are studied based on TGR and RR methods. The study also makes an effort to propose a novelty approach to improve these methods by means of FLA. Table 41 presents at a glance the number and gas concentrations of transformers studied in this chapter. The oil samples were collected from various power plants within South Africa.

Table 41. Gas concentration in test transformers

Trf. #	CH_4	H_2	C_2H_2	C_2H_6	C_2H_4	TCG
1	12462	1003	113	22349	61160	97087
2	6933	380	48	18968	46002	72331
3	26	2	2	8	6	44
4	438	1478	15	561	36	2528
5	261	1716	20	381	5	2383
6	314	1157	10	330	10	1821
7	1701	45	1	2	13	1762
8	106	487	0	217	10	820
9	27	89	72	423	16	627
10	594	4040	30	838	8	5510

4. RESULTS AND DISCUSSION

Based upon the DGA oil sample results, ten transformers were diagnosed using the RRA, IEC FLA and proposed FLA for comparison. It has been observed that with the proposed FLA, a higher precision on the diagnosis of incipient can be established. Table 42 presents the diagnostic findings of the tested transformers.

Table 42. Failure mechanism of test transformers

Trf. #	Failure Mechanism		
	RRA	IEC FL approach	Proposed FL Approach
1	Non diagnosable	Thermal fault of medium temp, 300-700°C	Excessive heating of leads to tank >900°C
2	Non diagnosable	Thermal fault of medium temp, 300-700°C	Excessive vibrations
3	Non diagnosable	Non diagnosable	Increase in temp. 150-200°C
4	Non diagnosable	Discharges of low energy	Surface tracking of cellulose insulation
5	Non diagnosable	Non diagnosable	Surface tracking of cellulose insulation
6	Non diagnosable	Non diagnosable	Surface tracking of cellulose insulation
7	Non diagnosable	Thermal fault of high temp, >700°C	Damage to A or C Phase HV winding
8	Non diagnosable	No Fault	Damage to Interturn
9	Non diagnosable	Non diagnosable	Damage to Interturn
10	Non diagnosable	Non diagnosable	Surface tracking of cellulose insulation

It is observed that the proposed FLA, in comparison to the RRA and IEC FLA, can detect incipient faults with some degree of authenticity. The results have also indicated some patterns that if the ratios are too small, the RRA and IEC FLA cannot quite diagnose the nature of the fault. On the other hand, the proposed FLA is handier in diagnosing the real fault.

A comparison of the three methods and corresponding gas ratios used in the fault diagnosis are presented in Table 43.

Table 43. Diagnostic methods comparison

Ratio/Method	RR	FL (IEC)	FL (PROPOSED)
C_2H_2/C_2H_4	√	√	-
CH_4/H_2	√	√	√
C_2H_4/C_2H_6	√	√	√
C_2H_6/CH_4	-	-	√
C_2H_2/C_2H_4	-	-	-
C_2H_6/C_2H_2	-	-	-

CONCLUSION

To appraise the robustness of the proposed FLA for diagnosing faults and level of severity, a large number of oil samples from a fleet of transformers of various sizes and service lifetimes has been introduced as master data in the formulation. DGA was conducted across all oil samples wherefrom dissolved gas concentrations in ppm was measured. Considering that it was challenging to surmise the imprecision of individual testing instruments and the factor of human error on each sample, a 95% confidence level was presumed to address these ambiguities. The output of the proposed FLA model introduced a gas ratio and revised dissolved gas ratio code range as indicated in Table 40. In addition, the proposed FLA introduce a novelty incipient fault diagnosis mechanism in Table 40 to close the gap of the shortcomings of RRA and the IEC FLA accommodate various dissolved gas ratios and corresponding fault mechanism which would then result in no diagnosis of the DGA oil sample results. The novelty methodology proposed herein is intended to be adequate to transformer manufacturers and utility owners in identifying incipient faults and effective planning and maintenance practices.

The authors intend to continuously improve the proposed incipient fault mechanism by conducting more DGA studies in future publications.

REFERENCES

[1] IEEE Guide for the Interpretation of Gases Generated in Oil-Immersed Transformers," in *IEEE Std C57.104-1991*, vol., no., pp.0_1-, 1992, doi: 10.1109/IEEESTD.1992.106973.

[2] IEEE Guide for the Interpretation of Gases Generated in Oil-Immersed Transformers," in *IEEE Std C57.104-2008* (Revision of IEEE Std C57.104-1991), vol., no., pp.1-36, 2 Feb. 2009, doi: 10.1109/IEEESTD.2009.4776518.

[3] IEEE Draft Guide for the Interpretation of Gases Generated in Oil-Immersed Transformers," in *IEEE PC57.104/D6*, November 2018, vol., no., pp.1-112, 2 Dec. 2018.

[4] IEEE Guide for the Interpretation of Gases Generated in Mineral Oil-Immersed Transformers," in *IEEE Std C57.104-2019* (Revision of IEEE Std C57.104-2008), vol., no., pp.1-98, 1 Nov. 2019, doi: 10.1109/IEEESTD.2019.8890040.

[5] Jane L. K., S. A Borakhade" Dissolved Gas Analysis in Transformer using Three Gas Ratio Method and Fuzzy Logic based on IEC Standard," *SSRG International Journal of Electrical and Electronics Engineering* (SSRG-IJEEE) – volume 2 Issue 4 April 2015.

[6] Liu L. Z., B. Song, E. Li, Y. Mao and G. Wang, "Study of "code absence" in the IEC three-ratio method of dissolved gas analysis," in *IEEE Electrical Insulation Magazine*, vol. 31, no. 6, pp. 6-12, November-December 2015, doi: 10.1109/MEI.2015.7303257.

[7] Gouda O. E., S. H. El-Hoshy and H. H. E.L.-Tamaly, "Proposed three ratios technique for the interpretation of mineral oil transformers based dissolved gas analysis," in *IET Generation, Transmission & Distribution*, vol. 12, no. 11, pp. 2650-2661, 19 6 2018, doi: 10.1049/iet-gtd.2017.1927.

[8] Ma H., Zheng Li, P. Ju, J. Han and L. Zhang, "Diagnosis of power transformer faults on fuzzy three-ratio method," *2005 International Power Engineering Conference*, Singapore, 2005, pp. 1-456, doi: 10.1109/IPEC.2005.206897.

[9] Taha I. B. M., S. S. M. Ghoneim and A. S. A. Duaywah, "Refining DGA methods of IEC Code and Rogers four ratios for transformer fault diagnosis," *2016 IEEE Power and Energy Society General Meeting* (PESGM), Boston, MA, 2016, pp. 1-5, doi: 10.1109/PESGM.2016.7741157.

[10] Zope N., S. I. Ali, S. Padmanaban, M. S. Bhaskar and L. Mihet-Popa, "Analysis of 132kV/33kV 15MVA power transformer dissolved gas using transport-X Kelman Kit through Duval's triangle and Roger's Ratio prediction," *2018 IEEE International Conference on Industrial Technology* (ICIT), Lyon, 2018, pp. 1160-1164, doi: 10.1109/ICIT.2018.8352342.

[11] Mohamad F., K. Hosny and T. Barakat, "Incipient Fault Detection of Electric Power Transformers Using Fuzzy Logic Based on Roger's and IEC Method," *2019 14th International Conference on Computer Engineering and Systems* (ICCES), Cairo, Egypt, 2019, pp. 303-309, doi: 10.1109/ICCES48960.2019.9068132.

[12] Nyandeni D. B, M. Phoshoko, R. Murray, B.A Thango, "Transformer Oil Degradation on PV Plants – A Case Study," *8th South African Regional Conference* (CIGRE), 14-17 November, 2017.

[13] Permana S, S Sumarto and W S Saputra," Analysis of Transformer Conditions using Triangle Duval Method," *International Symposium on Materials and Electrical Engineering* (ISMEE), Volume 384, conference 1, 2017.

[14] Sukhbir Singh and M. N. Bandyopadhyay, "Duval Triangle: A Noble Technique for DGA in Power Transformers" *International Journal of Electrical and Power Engineering*, 4: 193-197, 2010.

[15] Desouky S. S., A. E. Kalas, R. A. A. El-Aal and A. M. M. Hassan, "Modification of Duval triangle for diagnostic transformer fault through a procedure of dissolved gases analysis," *2016 IEEE 16th International Conference on Environment and Electrical Engineering* (EEEIC), Florence, 2016, pp. 1-5, doi: 10.1109/EEEIC.2016.7555796.

[16] Abdelaziz Lakehal and Fouad Tachi, "*Bayesian Duval Triangle Method for Fault Prediction and Assessment of Oil Immersed Transformers*," August 18, 2017. https://doi.org/10.1177/002 0294017707461.

[17] Irungu G. K., A. O. Akumu and J. L. Munda, "Comparison of IEC 60599 gas ratios and an integrated fuzzy-evidential reasoning approach in fault identification using dissolved gas analysis," *2016 51st International Universities Power Engineering Conference* (UPEC), Coimbra, 2016, pp. 1-6, doi: 10.1109/UPEC.2016.8114055.

[18] Abu-Siada A., S. Hmood and S. Islam, "A new fuzzy logic approach for consistent interpretation of dissolved gas-in-oil analysis," in *IEEE Transactions on Dielectrics and Electrical Insulation*, vol. 20, no. 6, pp. 2343-2349, December 2013, doi: 10.1109/TDEI.2013.6678888.

[19] Hmood S., A. Abu-Siada, M. A. S. Masoum and S. M. Islam, "Standardization of DGA interpretation techniques using fuzzy logic approach," 2012 *IEEE International Conference on Condition Monitoring and Diagnosis*, Bali, 2012, pp. 929-932, doi: 10.1109/CMD.2012.6416305.

[20] Wattakapaiboon W. and N. Pattanadech, "The state of the art for dissolved gas analysis based on interpretation techniques," *2016 International Conference on Condition Monitoring and Diagnosis* (CMD), Xi'an, 2016, pp. 60-63, doi: 10.1109/CMD.2016.7757763.

Chapter 8

LOSS FINANCIAL EVALUATION

In the recent multifaceted transformation dawn of low-carbon power generation markets and escalated growth of the renewable energy technologies in South Africa, a comprehensive analysis of transformers' technical and economic performance has been under-explored as a result of insufficient knowledge in this study area. The Transformer Total Ownership Cost (TTOC) is an economic evaluation pre-planned to give the transformer purchasers and owners the intended service-lifetime costs basis of their transformer's investment. The procedure for evaluating the transformer TTOC depend upon the concept of service-lifetime loss evaluation of transformers. Meticulously, this techno-economic concept considers the arithmetic sum of the presented discounted value (PDV) of a unit kilowatt loss of the transformer during its intended service lifetime. In a competitive bidding process, the TTOC is employed as a criterion to select the most suitable option among competing offers by transformer manufacturers and substantiate the purchasing decision of the units capable of achieving maximum productivity at minimum expense.

This chapter provides transformers' service-lifetime loss evaluation method, which fulfils some of the operational requirements that have been introduced from the recent deployment of solar photovoltaic plants into the modern South African energy mixed electrical network. The proposed method herein takes into account the intermittent nature of the solar PV, annualized energy cost and services losses. The results yield a TTOC optimized for the operational requirements of modern solar PV plant in a South African context.

1. Introduction

The most frequent practice employed by transformer purchasers and manufacturers for evaluating the cost efficacy of transformers is founded on the Transformer Total Ownership Cost (TTOC) method, in which the TTOC is defined as the arithmetic sum of the transformer purchase cost and the cost of the transformer service lifetime losses [1-8]. The TTOC method contemplates the distinctive economic conditions encountered by each power utility when making a transformer purchase decision. The changes in the cost of energy, loading and investment are described by way of two appraisal factors, referred to as the F_{NL} and F_{LL} service lifetime loss evaluation factors and corresponds to the cost of no-load and load losses, respectively [9–13]. It is critical to take notice that these factors should differ with respect to the role of the power utility in the South African energy mix. The TTOC method for a power utility like the Electricity Supply Commission (ESKOM) necessitates a complete knowledge and evaluation of the cost of transformer losses for generation, transmission and distribution, whereas the transformer service lifetime loss evaluation method for industrial application calls for comprehensive knowledge and estimation of the energy rate paid to the supplying power utility.

Transformers intended to operate within solar PVs will be subjected to the intermittent nature of the facility, as shown in Figure 18 below. On a 24-hours daily cycle, there are about 14 and 10 hours of high sunlight and low sunlight intensity, respectively, seen by the PV modules [14]. Consequently, the generating state factor (FGS) and non-generating state factor (FNGS) will be applicable to these transformers, which is the ratio of the generating and non-generating mode of the plant on the 24 hours daily may be established as 0.583 and 0.417 respectively.

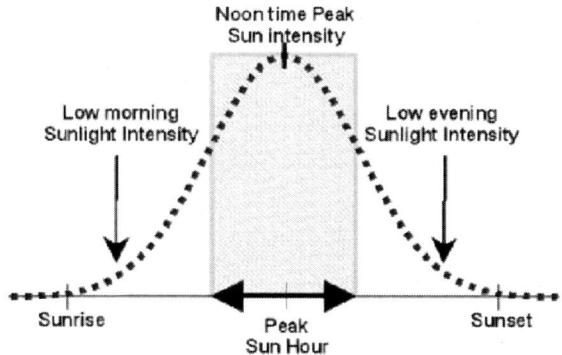

Figure 18. Solar PV generation profile.

This chapter is arranged as follow. Section 2 demonstrate the conventional TTOC method with an environmental cost. Section 3 illustrates the proposed TTOC methodology intended for large-scale solar plant application. Section 4 present a case study for offers by various transformer manufactures. Moreover, the proposed TTOC is applied and compared with the conventional TTOC method for each offering to determine the most economic evaluation of transformers for the LSSP. Section 5 discuss the results of this chapter. Section 6 describe the overview and section 7 concludes the chapter.

2. CONVENTIONAL TTOC MODEL WITH ENVIRONMENTAL COST

The most prevalent technique for evaluating the most economical performance of transformers is the TTOC method, which is founded on the formula as expressed in eq. (1) [15-17].

$$TTOC = P_C + C_L + C_e \tag{1}$$

Here, the TTOC present the transformer total ownership cost (in ZAR), the P_C make reference to the transformer purchasing cost (in ZAR) and C_L (in ZAR) allude to the transformer service losses over its intended service lifetime. Finally, C_e (in ZAR) [8],[9],[10] refers to the cost of emissions generally incurred on coal power stations in South Africa. It follows that there is calculated as expressed in eq. (2).

$$C_L = C_{NL} + C_{LL} \tag{2}$$

Here,

$$C_{NL} = F_{NL} \times P_{NL} \tag{3}$$

$$C_{LL} = F_{LL} \times P_{LL} \tag{4}$$

where, C_{NL} and C_{LL} make reference to the cost of the transformer no-load (P_{NL}) and load losses (P_{LL}) respectively. Additionally, F_{NL} and F_{LL} ascribe to the transformer no-load (in ZAR/kW) and load loss factors (in ZAR/kW), respectively. By integrating eq. (1) - eq. (4), the conventional TTOC formula is established as expressed in eq. (5).

$$TTOC = P_C + F_{NL} \times P_{NL} + F_{LL} \times P_{LL} \tag{5}$$

The factors F_{NL} and F_{LL} are calculated in accordance with eq. (6) and eq. (7) respectively.

$$F_{NL} = \frac{(1+i)^n - 1}{i.(1+i)^n} \times E_{kWh} \times 8760 \tag{6}$$

The no-load loss factor in a 25 year period is shown in Figure 19.

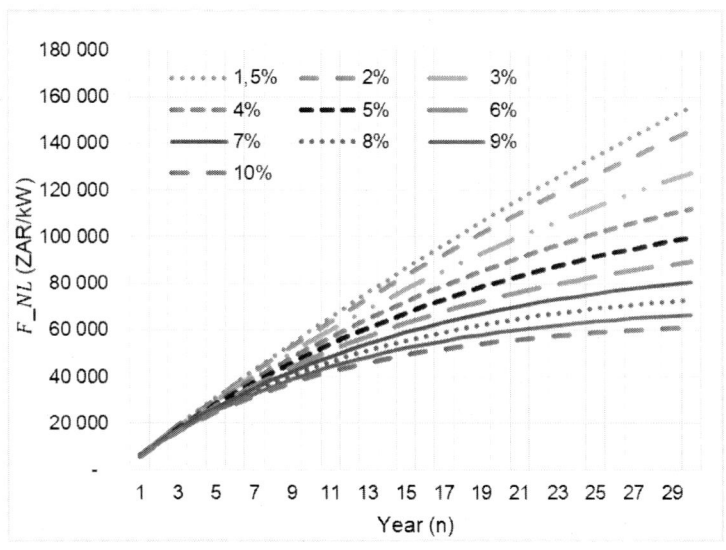

Figure 19. No-load Capitalization Factor.

$$F_{LL} = \frac{(1+i)^n - 1}{i.(1+i)^n} \times E_{kWh} \times 8760 \times \left(\frac{I_1}{I_r}\right)^2 \tag{7}$$

Here,
i – Interest Rate [%/Year]
n – Service lifetime [Years]
E_{kWh} – Cost of energy [ZAR/kWh]
8760 – Number of hours in operation [Hr./Year]
I_1 – Transformer loading current in Amps [A]
I_r – Transformer rated current in Amps [A].

Table 44. Typical Transformer load factors

Rating (kVA)	Load Factor, I_1/I_r
≤ 200	0.35
≤ 200	0.4
>315	0.65

The load-loss capitalization factor using Eq. (6) and considering an interest rate of 1.5% yields and an energy tariff of about 0.75 ZAR/kWh. The load loss factor in a 25 year period is shown in Figure 20.

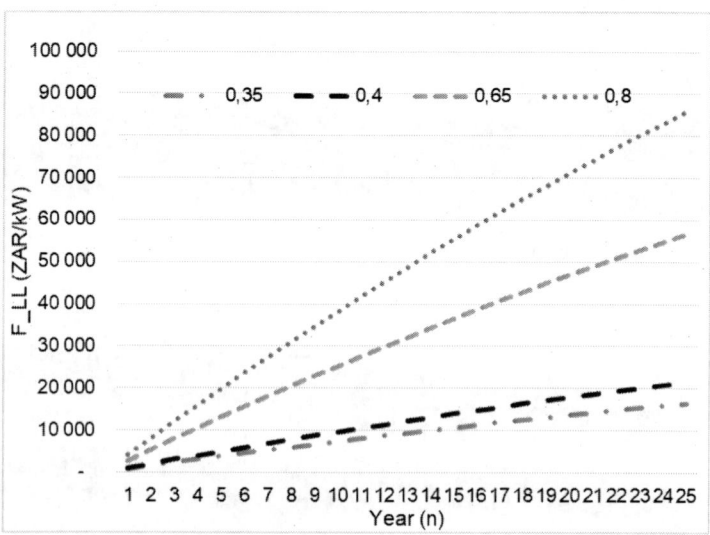

Figure 20. Load-loss Capitalization Factor.

Generally, the F_{NL} and F_{LL} service lifetime loss evaluation factors form part of the technical specifications submitted by the IPP's to the transformer manufacturers when a tender is issued, who are then challenged to optimize their designs to give the required operational performance. The outcome of applying this method in evaluating various transformer offers should be the most economic transformer with low TTOC optimized for the specified operating conditions.

3. CALCULATION OF THE F_{NL} AND F_{LL} FACTORS WITH INTERMITTENT GENERATION PROFILE

It can be well established that the F_{NL} and F_{LL} service lifetime loss evaluation factors presented in this section are appropriate to advance the collective interests of Independent Power Producers in South Africa that, in partnership with Eskom, supply power to the grid by way of step-up transformers to give a sustainable and complementary solution to the South African power generation requirements. The keystone in the capitalization of transformer losses is a comprehensive definition of the energy and demand elements of the cost of the transformer losses, C_L. The energy element is the present electricity cost per unit (in ZAR/kWh) that will be consumed by each kilowatt of transformer loss over its intended service lifetime. Additionally, the demand element is the expenses incurred by the plant to back up the power consumed by the transformer losses. To this point, the proposed method will acknowledge entirely by what means these elements should be assessed.

In the course of the day, to a great extent the large-scale solar plant is expected to change its generation conditions owing to the availability of solar irradiation. When the plant is its Generating Mode (GM), it is completely accountable for its energy and loss requirements and supply of energy to the electrical grid. In the case of Non-Generating Mode (NGM), the supply of complementary operations of the plant and losses will be provided by the main electrical grid. In essence, the IPP will purchase power from power ESKOM when the generation capacity is minimum.

$$TTOC = P_C + C_L \tag{8}$$

The cost of the transformer losses (in ZAR) for a large-scale solar plant ($C_{L(LSSP)}$) is presented in eq. (9). This proposed method considers the arithmetic sum of the present cost of each electrical energy (in ZAR) of the transformer no-load, load and auxiliary service loss throughout its intended lifetime.

$$C_{L(LSSP)} = C_{NL} + C_{LL} + C_A \tag{9}$$

In eq. (10) - (12), three-loss components viz. the no-load, load and auxiliary service loss are considered in the computation of the cost of each loss component.

$$C_{NL} = F_{NL} \times P_{NL} \tag{10}$$

$$C_{LL} = F_{LL} \times P_{LL} \tag{11}$$

$$C_A = F_A \times P_{LL} \tag{12}$$

Correspondingly, Eq. (13) - (15) demonstrate the foundational principle of transformer loss financial evaluation method that acknowledge the operational requirements of a LSSP. It is entirely founded upon the transformer service lifetime no-load and load loss components, as well as the plant's GM and NGM. In eq. (13), the no-load loss will be capitalized under both GM and NGM by considering the cost of electricity for both the solar plant and electrical utility. The annuitized rates considered in eq. (13) is the cost of electricity for the energy supplied by the grid $E_{kWh(grid)}$ and solar power plant $E_{kWh(PV)}$ the transformer will consume that during the service lifetime.

$$F_{NL} = \frac{(1+i)^n - 1}{i.(1+i)^n} \times 8760 \times \left(E_{kWh(grid)} \times F_{NGM} + E_{kWh(PV)} \times F_{GM}\right)$$
(13)

The load and auxiliary losses in Eq. (14)-(15) are determined only during the plant's GM due to the fact that these losses will be more pronounced during this condition. In the case on the no-load losses, they will eventuate when the transformer is energized at both the plant's GM and NGM.

$$F_{LL} = \frac{(1+i)^n - 1}{i.(1+i)^n} \times E_{kWh(PV)} \times 8760 \times F_{GM} \times \left(\frac{I_1}{I_r}\right)^2 \quad (14)$$

$$F_A = E_{kWh(PV)} \times 8760 \times F_{GM} \quad (15)$$

4. CASE STUDY

In competitive bidding, the proposed TTOC methodology is applied in the economic performance evaluation of oil-immersed step-up transformers of three different transformer manufactures, as presented in Table 45. It can be observed that manufacturer A has the cheapest P_C and comprise of the highest total losses, whilst offer C has the lowest total losses and highest P_C.

Table 45. Transformer manufacturer offers

Offer	P_C, ZAR	P_{NL}, kW	P_{LL}, kW
A	550 000	0.695	2,8
B	681 750	0.635	2,55
C	800 000	0.595	2,4

The energy tariffs for the coal power generation that feeds the solar PV during non-generating states ($E_{kWh(grid)}$) and the solar PV tariff ($E_{kWh(PV)}$) that facilitate the plant during the generating state are presented in the figure below. The classical ACOE is calculated based on the assumption that the unit will be in operation for 8760 per annum with an energy cost of 0.74 ZAR/kWh and inflation of 1.5% per annum. For the solar PV ACOE, an energy cost of 0.58 ZAR/kWh, non-generating and generating factors are 0.583 and 0.417 respectively are considered. The annualized cost of energy over 20 year period is shown in Figure 21.

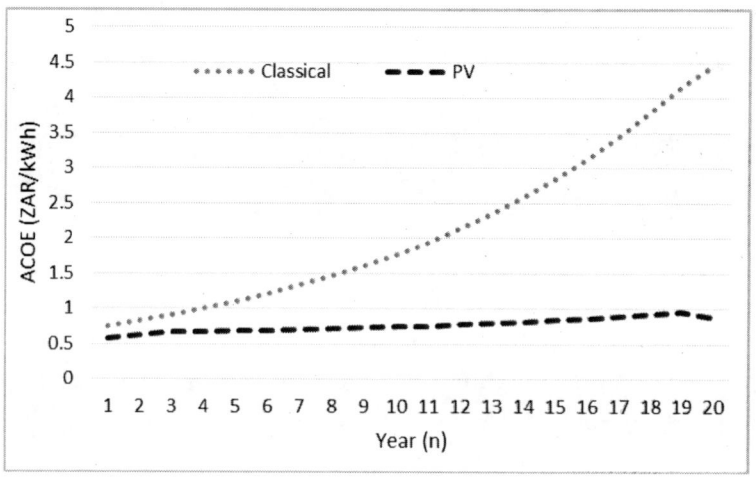

Figure 21. Annualized cost of energy over 20 year period.

Figure 26 shows the capitalization factor of the no-load graphed for a 20 years period that the transformer under study in service.

In order to determine the F_{NL} and F_{LL} service lifetime loss evaluation factors, the data presented in Table 46 is pivotal. The results of the classical method are tabulated below for each of the transformer offers.

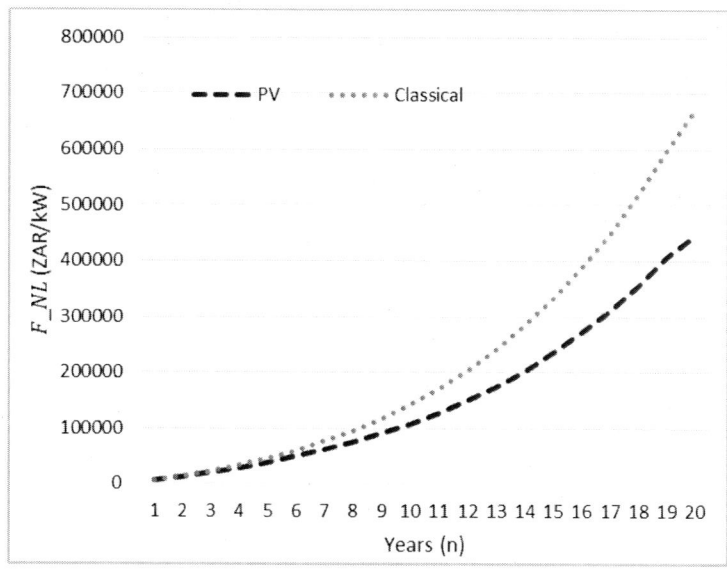

Figure 22. Comparison of F_{NL} for the solar PV.

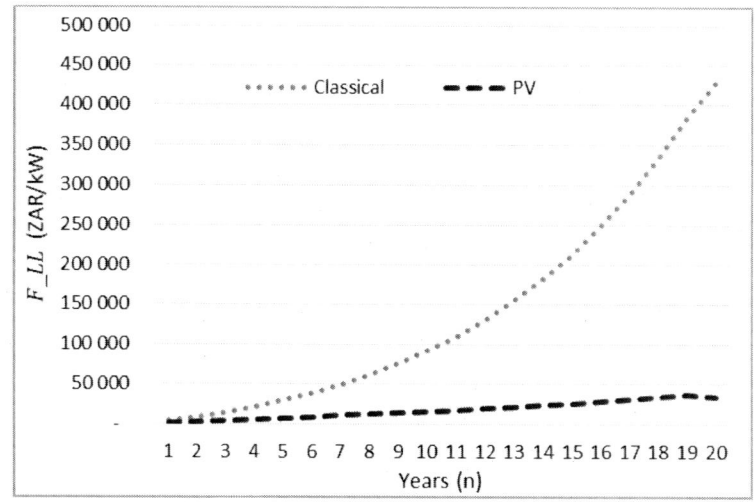

Figure 23. Comparison of F_{LL} for the solar PV.

Table 46. Cost of Service losses (Classical)

Year	A	B	C
1	16 098	14 655	13 794
5	114 414	104 155	98 036
10	355 219	323 368	304 371
15	827 906	753 671	709 396
20	1 667 087	1 517 606	1 428 453

The results indicate that over a 20 year period, offer C will have the least total cost of services losses while offer A will have the highest cost of losses. The results based on the solar PV method are also presented in Table 47 below. It can be observed that the cost of service losses for offer C is 24% lower than the classical method. These results substantiate that the classical method cannot capture the actual operational requirements of transformers facilitating solar PV environments.

Table 47. Cost of Service losses (PV)

Year	A	B	C
1	7 814	7 111	6 694
5	48 179	43 840	41 268
10	119 955	109 142	102 741
15	236 588	215 241	202 622
20	406 989	370 229	348 530

Table 48 shows the TOC for the offers optimized to the classical loss evaluations at 75°C reference temperature.

The data above shows that transformers intended for solar PV applications are slightly more expensive than conventional distribution transformers. Offer C has the most optimal losses but a slightly higher cost than offer A, which has higher losses. A similar trend is observed in Table 49 below.

Table 48. TTOC losses (Classical)

Year	A	B	C
1	566 098	696 405	813 794
5	664 414	785 905	898 036
10	905 219	1 005 118	1 104 371
15	1 377 906	1 435 421	1 509 396
20	2 217 087	2 199 356	2 228 453

Table 49. TTOC (PV)

Year	A	B	C
1	557 814	688 861	806 694
5	598 179	725 590	841 268
10	669 955	790 892	902 741
15	786 588	896 991	1 002 622
20	956 989	1 051 979	1 148 530

The TTOC of the proposed solar PV method is 52% lower than the classical method. In the purchase decision, the utility owner will make a good investment by selecting offer C. during service, solar PV transformers are susceptible to harmonic current contents which cause additional losses, and as a result low loss units are more suitable for this application. Over the 20 years, offer A is more likely to generate oil stray gassing phenomenal as a results of additional losses and hotspot temperature rise.

5. RESULTS AND DISCUSSION

It can be well established that the F_{NL} and F_{LL} service lifetime loss evaluation the proposed TTOC method is critical not only for the transformer service lifetime management but also for engineers in the design or purchasing environment as a result of the below-mentioned findings:

- The transformer with the cheaper price initially has higher losses. However, it has the minimum cost of service losses by the end of the 20 year period, as observed in Table 46 and Table 47.
- It is observed from the results that optimal loss selection of the transformer based on the total ownership cost method offers the most economic benefits over the reminder of the transformer lifetime from the point of purchase. Consequently, electrical designers can minimize the transformer total ownership cost by further optimizing the transformer design.
- Purchasing engineers can apply the proposed TTOC in the selection of transformer offers, and they can reap the reward of determining the economic benefits of selecting high and low initial purchasing costs, a high and low loss against each other, respectively.
- The IPP's can profit from the proposed TTOC so as to minimize energy costs and improve the electrical power system efficiency by selecting, installing and commissioning to a great extent energy-efficient transformers.
- The proposed methods for evaluating the capitalization factors captured operational requirements that the classical method is not considering.

CONCLUSION

This chapter proposed an inventive method for computing the capitalization factors for transformers by introducing the intermittent generation profile characteristics of a large-scale solar plant, energy tariffs and service losses into the classical service loss evaluation technique. The proposed techniques were applied for the financial evaluation of transformers intended to operate in a solar PV plant using actual financial and technical data. The results are compared against the conventional TTOC method to indicate the importance of integrating a large-scale solar plant's intermittent generation profile characteristics.

REFERENCES

[1] Indarto A., I. Garniwa, R. Setiabudy and C. Hudaya, "Total cost of ownership analysis of 60 MVA 150/120 kV power transformer," *2017 15th International Conference on Quality in Research (QiR): International Symposium on Electrical and Computer Engineering*, Nusa Dua, Bali, Indonesia, 2017, pp. 291-295, doi: 10.1109/QIR.2017.8168499.

[2] Knutson T., "Conducting Distribution Transformer Evaluations Using the Total Ownership Cost Method," *2015 IEEE Rural Electric Power Conference*, Asheville, NC, USA, 2015, pp. 97-101, doi: 10.1109/REPC.2015.14.

[3] Thango B. A., J. A. Jordaan and A. F. nnachi, "Total Ownership Cost Evaluation for Transformers within Solar Power Plants," *2020 6th IEEE International Energy Conference* (ENERGYCon), Gammarth, Tunisia, 2020, pp. 302-307, doi: 10.1109/ENERGYCon48941.2020.9236613.

[4] Lazari A. L. and C. A. Charalambous, "Probabilistic Total Ownership Cost of Power Transformers Serving Large-Scale Wind Plants in Liberalized Electricity Markets," in *IEEE Transactions on Power Delivery*, vol. 30, no. 4, pp. 1923-1930, Aug. 2015, doi: 10.1109/TPWRD.2014.2365832.

[5] Guo-Hua Q., R. Zheng, S. Lei, Z. Bo, X. Jian-Gang and Z. Xiang-Ling, "A new life cycle cost model of power transformer and its comprehensive sensitivity analysis," *2014 International Conference on Power System Technology*, Chengdu, China, 2014, pp. 1342-1348, doi: 10.1109/POWERCON.2014.6993659.

[6] Lee S. H., A. K. Lee and J. O. Kim, "Determining economic life cycle for power transformer based on life cycle cost analysis," *2012 IEEE International Power Modulator and High Voltage Conference* (IPMHVC), San Diego, CA, USA, 2012, pp. 604-607, doi: 10.1109/IPMHVC.2012.6518816.

[7] Zhang Y., J. Jiao, Y. Yang, X. Li, T. Zhou and X. Huang, "A new maintenance decision making model based on life cycle cost analysis for power transformers," *2016 IEEE International Conference on High Voltage Engineering and Application* (ICHVE), Chengdu, China, 2016, pp. 1-4, doi: 10.1109/ICHVE.2016.7800855.

[8] Bian J., S. Yang and X. Sun, "The optimal maintenance strategy of power transformers based on the life cycle cost," 2017 *13th International Conference on Natural Computation, Fuzzy Systems and Knowledge Discovery* (ICNC-FSKD), Guilin, China, 2017, pp. 2354-2358, doi: 10.1109/FSKD.2017.8393140.

[9] Lazari A. L. and C. A. Charalambous, "Integrating fossil fuel mix and pricing in evaluating the Total Ownership Cost of distribution transformers of vertically integrated utilities," *2014 IEEE International Energy Conference* (ENERGYCON), Cavtat, Croatia, 2014, pp. 1184-1189, doi: 10.1109/ENERGYCON.2014.6850573.

[10] Charalambous C. A., A. Milidonis, A. Lazari and A. I. Nikolaidis, "Loss Evaluation and Total Ownership Cost of Power Transformers—Part I: A Comprehensive Method," in *IEEE Transactions on Power Delivery*, vol. 28, no. 3, pp. 1872-1880, July 2013, doi: 10.1109/TPWRD.2013.2262506.

[11] Chowdhury A. A., L. Bertling and D. E. Custer, "Determining Distribution Substation Transformer Optimal Loadings Using a Reliability Cost-Benefit Approach," *2006 International Conference on Probabilistic Methods Applied to Power Systems*, Stockholm, Sweden, 2006, pp. 1-9, doi: 10.1109/PMAPS.2006.360426.

[12] Wen H., H. Zou, C. Ouyang, X. Bao, H. Li and X. Li, "Exploration on energy-saving effect of amorphous transformers extended in Nanning based on theoretical analysis and Total Owning Cost methods (TOC)," *2008 China International Conference on Electricity Distribution*, Guangzhou, China, 2008, pp. 1-8, doi: 10.1109/CICED.2008.5211678.

[13] "IEEE Guide for Loss Evaluation of Distribution and Power Transformers and Reactors," in *IEEE Std C57.120-2017* (Revision of IEEE Std C57.120-1991), vol., no., pp.1-53, 18 Oct. 2017, doi: 10.1109/IEEESTD.2017.8103991.

[14] Suechoey B., J. Ekburanaway, N. Kraisnachinda, S. Banjongjit and M. Kando, "An analysis and selection of distribution transformer for losses reduction," *2000 IEEE Power Engineering Society Winter Meeting. Conference Proceedings* (Cat. No.00CH37077), Singapore, 2000, pp. 2290-2293 vol.3, doi: 10.1109/PESW.2000.847711.

[15] Thango B. A., K. Moloi, J. A. Jordaan and A. F. Nnnach, "A Further Look into the Service Lifetime Cost of Solar Photovoltaic Energy Transformers," *2021 Southern African Universities Power Engineering Conference/Robotics and Mechatronics/Pattern Recognition Association of South Africa (SAUPEC/RobMech/PRASA)*, Potchefstroom, South Africa, 2021, pp. 1-7, doi: 10.1109/SAUPEC/RobMech/PRASA52254.2021.9377229.

ABOUT THE AUTHORS

Bonginkosi A. Thango obtained his Bachelor of Technology (BTech) in Electrical Engineering in 2017 at the Tshwane University of Technology, South Africa. In 2019, he was awarded Cum Laude for Master of Engineering (MEng) in the same University. He is currently pursuing his Doctor of Engineering (DEng) degree with the same University. His research interest includes Finite Element Method modelling, condition monitoring of transformers, Renewable Energy, Data Analysis and Mathematical modelling.

About the Authors

Jacobus A. Jordaan obtained his Bachelor's degree in Electrical/Electronic Engineering (BEng) in 1997 from the North-west University, South Africa, and his MEng degree in 2002 from the same university. He then obtained a double master's degree in Electronic Engineering and Engineering Management in 2004 from the South Westphalia University of Applied Sciences (SOEST), Germany and University of Bolton. In 2007 He obtained a doctorate degree in Electrical Engineering at the Tshwane University of Technology, South Africa. He is currently an associate professor at the same university, with his interests and field of research including microcontrollers, programming, signal processing and mathematical modelling.

Agha F. Nnachi obtained his Bachelor degree (BEng.) in electrical engineering in 2002 at Nnamdi Azikiwe University Awka, Nigeria. In 2009/2010, he received double master's degree MTech and MSc in electrical engineering at Tshwane University of Technology. In 2015, he obtained government certificate of competency (GCC) in electrical engineering. In 2017, he received Doctorate degree in electrical engineering at Tshwane University of Technology. He is a registered professional engineer with Engineering Council of South Africa and a senior member of the South African Institute of Electrical Engineers. His research interest includes power electronic converters, high voltage DC, renewable energy and power quality.

Leon S. Sikhosana obtained his double Diploma in Electrical Engineering for both heavy and light current in 2017 at the Tshwane University of Technology, South Africa. In 2020, he was awarded for completing his Bachelor of Technology Engineering (BTech) in the same University. He is aiming to pursue his Master of Engineering (M.Eng.) degree with the same University. His research interest includes transformers in Renewable Energy applications.

Aloys O. Akumu graduated with an MSc degree in Electrical Engineering in 1994 from Poznan Technical University, Poland, and a DEng degree in Electrical Engineering in 2003 from Ehime University, Japan. He is currently a Senior Lecturer at the Tshwane University of Technology, South Africa. His interest is in the condition monitoring of power transformers, artificial intelligence, and smart grids.

Bolanle T. Abe is currently the Acting Academic Manager of the Department of Electrical Engineering, Tshwane University of Technology, South Africa. In 2003, she earned her Master of Electrical Engineering degree (M.Eng) at the Federal University of Technology, Akure, Nigeria. Bolanle later proceeded for her PhD degree at the School of Electrical and Information Engineering, University of the Witwatersrand, Johannesburg, South Africa. In 2014, she obtained her Doctor of Philosophy (PhD) in Electrical Engineering at the University. Dr Abe is a member of the South African Institute of Electrical Engineering. Her research activities include Electromagnetic Compatibility, Communication Engineering, Machine Learning and Remote Sensing applications.

Udochukwu B. Akuru obtained the PhD degree in Electrical Engineering from Stellenbosch University, South Africa in December 2017, and both MEng and BEng degrees from University of Nigeria,

Nsukka, in June 2013 and August 2008, respectively. He is currently a Senior Lecturer at Tshwane University of Technology, South Africa. He held a Postdoctoral Research Fellowship in Stellenbosch University, South Africa, between 2018 and 2020. He is also affiliated with the University of Nigeria, Nsukka, since August 2011. He is a registered engineer, Senior Member of IEEE and SAIEE as well as NRF rated researcher in South Africa. His main research interests are in electrical machines design and renewable energy technologies.

INDEX

A

ANSI, 44

C

cellulose insulation, v, 73, 74, 84, 116, 118
copper losses, 19, 34, 42, 46, 66
core clamp, 5, 22, 23, 42, 58, 60

D

DPVP transformer, 57, 62, 63, 65, 66, 69

F

factor-K, 45
flitch plate loss, 23
frame loss, 22
fuzzy logic, 109, 111, 113, 114, 120, 121, 122, 123

G

gas concentration, 93, 94, 95, 96, 97, 99, 100, 101, 102, 103, 110, 117, 119
greenhouse gases, vii, 2

H

H2 method, 4
harmonic conditions, ix, 2, 4, 5, 7, 19, 33, 44, 55, 60, 61
harmonic currents, ix, 2, 33, 40, 51, 107
harmonic effects, 42
harmonic load currents, v, 1, 41, 43
hotspot temperature, 3, 4, 5, 17, 19, 24, 30, 39, 55, 57, 58, 61, 62, 65, 69, 99, 103, 105, 137

I

IEEE, 15, 19, 30, 31, 32, 33, 34, 35, 36, 44, 51, 52, 53, 70, 71, 74, 85, 86, 87, 88, 89, 101, 102, 103, 105, 106, 107, 108, 120, 121, 122, 123, 139, 140, 141, 147

Index

inverters, ix, 3, 39, 57

K

K-factor, 37, 40, 41, 44, 47, 48, 49, 50, 53
K-rated transformers, v, 37, 41

L

load current, 2, 4, 5, 6, 7, 10, 11, 12, 13, 15, 19, 42, 43, 44, 45, 47, 48, 53, 55, 57, 58, 63, 64, 66, 67, 68, 69, 71
load-losses, 19

M

maximum load current, 6

O

oil-filled solar PV transformer, 1, 9

P

polymerization, v, 73, 81, 82, 107

R

renewable energy, vii, viii, 2, 13, 17, 18, 29, 53, 56, 57, 70, 125, 143, 145, 147
renewable energy applications, vii, viii, 18, 57, 70
renewable energy market, vii, 56
Roger ratio approach, 112

S

South Africa, vii, 1, 2, 13, 14, 23, 33, 38, 39, 51, 52, 55, 56, 57, 62, 69, 70, 71, 73, 92, 94, 97, 105, 106, 107, 117, 121, 125, 126, 128, 131, 141, 143, 144, 145, 146

T

tank wall loss, 21
transformer engineering, vii
transformer harmonic loss, 4, 10, 31
transformer losses, v, 1, 10, 25, 28, 30, 41, 42, 58, 61, 67, 68, 126, 131, 132
transformer service lifetime, ix, 126, 132, 137
transformers, v, vii, viii, ix, 1, 3, 4, 14, 15, 17, 18, 19, 21, 24, 25, 29, 30, 31, 32, 33, 34, 35, 36, 38, 39, 40, 41, 42, 43, 44, 51, 52, 53, 55, 57, 59, 70, 71, 73, 74, 75, 76, 77, 80, 81, 82, 83, 84, 85, 86, 87, 88, 89, 91, 92, 93, 94, 96, 97, 98, 99, 104, 105, 106, 107, 108, 109, 110, 112, 113, 114, 115, 117, 118, 119, 120, 121, 122, 125, 126, 127, 128, 131, 133, 136, 137, 138, 139, 140, 141, 143, 145, 146

W

winding Eddy loss, 1, 2, 3, 4, 5, 7, 11, 13, 17, 20, 21, 26, 27, 29, 30, 37, 42, 45, 46, 48, 50, 64, 66, 68